ANIMAL SCIENCE, ISSUES AND PROFESSIONS

WHALES AND DOLPHINS: BEHAVIOR, BIOLOGY AND DISTRIBUTION

ANIMAL SCIENCE, ISSUES AND PROFESSIONS

Additional books in this series can be found on Nova's website under the Series tab.

Additional E-books in this series can be found on Nova's website under the E-book tab.

ANIMAL SCIENCE, ISSUES AND PROFESSIONS

WHALES AND DOLPHINS: BEHAVIOR, BIOLOGY AND DISTRIBUTION

CRAIG A. MURRAY
EDITOR

Nova Science Publishers, Inc.
New York

For permission to use material from this book please contact us:
Telephone 631-231-7269; Fax 631-231-8175
Web Site: http://www.novapublishers.com

LIBRARY OF CONGRESS CATALOGING-IN-PUBLICATION DATA

Whales and dolphins : behavior, biology, and distribution / editor, Craig A. Murray.
p. cm.
Includes index.
ISBN 978-1-61668-693-2 (hardcover)
1. Cetacea. 2. Whales--Behavior. 3. Dolphins--Behavior. 4. Whales--Geographical distribution. 5. Dolphins--Geographical distribution. Murray, Craig A.
QL737.C4W4414 2010
599.5--dc22 2010007610

Published by Nova Science Publishers, Inc. ✢ New York

CONTENTS

PREFACE

The biology of cetaceans is one of the most compelling areas of research because of the extreme adaptations whales and dolphins have had to undergo in order to manage a life in the water. The fossil record of cetaceans is rich, and though much attention has been given to the origins of whales from terrestrial artiodactyls, it is important to realize that the biology, physiology, and behavior of modern cetaceans has not remained unchanged since this initial transition to being aquatic. This books discusses and presents new data on the behavior and biology of whales and dolphins including: the cenozoic environmental changes and evolution of baleen whales, the ecological and evolutionary divergence in whales and dolphins, the parasite fauna of cetaceans, and others.

Chapter 1- The earliest baleen-bearing whales (Mammalia, Cetacea, Mysticeti) are represented by fossils found in the Oligocene sediments of New Zealand, Japan, North America, and Europe. Their geographic distribution suggests that the origin of this particular mysticete group occurred in the remaining of the ancient Tethys ocean. During Miocene and Pliocene, baleen-bearing mysticetes dispersed all around the world and gave rise to several evolutionary radiations which are now identified at a taxonomic level in family-rank clades. These are Balaenidae, Balaenopteridae, Cetotheriidae, Eschrichtiidae, and Neobalaenidae. Some taxa are still to be considered incertae sedis and include archaic mysticetes from the Early and Middle Miocene of Japan, United States, Belgium, and Italy. The evolutionary radiations of baleen-bearing mysticetes are related to the development of particular characters linked to highly specialized feeding behaviors. In their turn, these are related to availability of particular kinds of food (i.e., krill, small-sized fish). It is clearly demonstrated that the abundance of krill in particular areas depends upon dynamic characters of oceanic currents, upwelling and other physical features of particular areas. For this reason, it should be possible to find a link between the evolutionary radiations of baleen-bearing whales and the environmental changes occurred in the oceans since their origin in the Late Oligocene time. The goals of this chapter are (a) to review the evolutionary radiations of baleen-bearing mysticetes; and (b) to individuate eventual environmental factors able to influence differential survival of peculiar mysticete forms in particular periods. The final synthesis will provide new ecological hypotheses explaining the increase/decrease of diversity in the group and, eventually, particular pattern of extinction and suggestions for conservation strategies of the living mysticete species.

Chapter 2- Near opposite poles on earth in their respective summers, an amazing process and journey occurs that has happened for thousands of year. Whales travel through icy waters, seeking copious amounts of food to nourish their massive bodies which are uniquely suited for their extreme environments.

Following the Arctic ice edge, navigating under thick ice or surfacing in open water or polynyas[1], the bowhead whale makes its annual journey similar to its ancestors countless generations before. It traces the path to the eastern Beaufort Sea where the rich waters team with food to nourish this thick skulled baleen whale. Constantly eating as it slowly swims, the bowhead skims the water for its desired meal of plankton and tiny crustraceans like krill, copepods and pteropods using its long and fine baleen filters.

In the Southern Ocean during the Antarctic summer, the blue whale lunges into dense groups of tiny krill, copepods and plankton. The planet's largest animal gulps huge mouthfuls by the ton as its throat pleats expand to form a gigantic food-laden pouch and then it forces water through hanging baleen plates which act as a sieve to catch the bounty.

The process of these large animals eating some of the smallest, occurs when their ecosystems are in balance. In simplest terms, for abundant food availability, the system needs the appropriate mix of sea ice, water temperature, currents, salinity, and sun light. These systems and the whales themselves are in serious danger and the culprit originates far from their pristine waters. Global climate change has begun to alter their habitats and those of whales in more temperate regions with documented adverse impacts

---

1 Polynyas are open water locations within the ice pack that are almost always clear of ice, caused by persistent physical conditions such as upwellings or wind patterns. They are important areas where marine mammals can surface to breathe, and can also provide productive feeding areas for bowheads, belugas, and narwhals. (Laidre & Heide-Jørgensen 2005).

already seen. What exact changes will occur is not certain but the outlook is especially grim unless substantial reductions in greenhouse gas emissions happen in short order.

Chapter 3- Speciation is the process by which populations of an ancestral species are subject to divergent selection that leads to genetic and phenotypic differences and reproductive isolation between populations and ultimately the formation of new species. Ecology has long been considered as having an integral role in this process as the major force behind divergent selection (Mayr 1942; Dobzhansky 1951; Endler 1977). Theoretical approaches to modeling this process of speciation across an ecological gradient (e.g. Dieckmann and Doebeli 1999; Doebeli and Dieckmann 2003) have been supported by empirical evidence of ecology's role in speciation or population differentiation (Funk et al. 2005; Schluter 2009). Classical theories suggested that this process could only occur in allopatry, i.e. in populations with non-overlapping geographic ranges, as extrinsic barriers separating populations would be necessary to stop gene flow and lead to reproductive isolation and the formation of new species (Mayr 1942). More recent theoretical models have shown how speciation can occur across an ecological gradient in parapatry (neighboring ranges) or sympatry (overlapping ranges) (Dieckmann and Doebeli 1999). A growing number of studies have identified genetic differentiation in widespread, continuously distributed highly mobile species, in the absence of geographic barriers, e.g. the common raven *Corvus corax* Omland et al. 2000; Giraffe *Giraffa camelopardalis* Brown et al. 2007 and Grey Wolf *Canis lupus* Musiani et al. 2007. Ecological factors have been increasingly explored as an intrinsic cause of inter- and intra-specific genetic differentiation in these studies (e.g. Musiani et al. 2007; Wolf et al. 2008).

The marine habitat has few of the extrinsic barriers that may cause isolation between populations (Palumbi 1994). Additionally marine species have a relatively low cost of locomotion (Tucker 1975; Williams 1999). Many marine species therefore have large home ranges e.g. blue-fin tuna Thunnus thynnus, great white shark *Carcharodon carcharias*, and leatherback turtles *Dermochelys coriacea* (Boustany et al. 2002; Block et al. 2005; Bonfil et al. 2005; Shillinger et al. 2008). Yet despite the lack of barriers to movement, many marine species show a high degree of population structuring (e.g. Bowen et al. 1992; Pardini et al. 2001; Carlsson et al. 2004). Site fidelity to breeding or feeding grounds appears to be an important factor in determining population structure (Baker et al. 1990; Palsbøll et al. 1995; Pardini et al. 2001). However, foraging preferences can lead to differentiation across less well-defined geographic boundaries and result in resource polymorphism

(Smith and Skúlason 1996). A number of well-studied whale and dolphin species appear to typify this (Hoelzel 2002).

Chapter 4- Cetaceans harbor a unique and apparently little diversified fauna of parasites (including commensals). Unlike their hosts, none of these whales' dwellers currently benefit from any conservation measure, perhaps because they are unappealing to humans, often cause harm to their hosts, and many can easily go unnoticed due to of their small size. However, we can put forward a convincing point for promoting conservation of these creatures: parasites can provide a wealth of insight into the evolutionary and demographic history of whales, two aspects of prime importance for conservation. For instance, genetic information of whale-lice has recently been used to reconstruct with great accuracy the population histories of the highly endangered species of right whales. Likewise, our preliminary study of the dolphin-dweller barnacle *Xenobalanus globicipitis* suggests that Mediterranean populations are strongly impacted by striped-dolphin die-offs. If this were further confirmed, it would open the possibility of tracking the history of high dolphin mortality events by using molecular markers of the barnacles. Parasites have also the potential to reconstruct ecological scenarios of cetaceans. For instance, recent evidence shows that a small group of tetraphyllidean cestodes use cetaceans as intermediate hosts and large sharks as definitive hosts. These parasites might, therefore, be useful to investigate the origin of the trophic interactions between sharks and cetaceans. Convincing as the above may seem, we could additionally argue that the parasites of cetaceans, like any other organism, are masterpieces of evolution that deserve equal conservation 'rights' as their hosts, hence the current concept of co-extinction risk. Indeed some of them represent fascinating examples of evolutionary novelty (e.g., barnacles settling on living tissues or copepods apparently able to feed on the keratin of baleen plates). From this holistic perspective, how could we scientists help preserving both cetaceans and their parasites? Among other measures, parasites should be sampled whenever the opportunity arises (e.g., from stranded cetaceans). Also, sampling protocols and research studies should (i) consider parasites beyond their pathogenic impact; (ii) be conceived within collaborative networks, and (iii) be adapted to the resource levels and time available of the teams involved.

Chapter 5- The biology of cetaceans is one of the most compelling because of the extreme adaptations whales and dolphins have had to evolve to manage a life in water. The fossil record of cetaceans is rich, and though much attention has been given to the origins of whales from terrestrial artiodactyls, it is important to realize that the biology, physiology, and behavior of modern cetaceans has not remained unchanged since this initial transition to being aquatic. Here I review some examples of how the fossil record of cetaceans informs us of how the evolution of anatomy, physiology and behavior has diverged and converged between and within the Odontoceti and Mysticeti in ways that would not be known if one were only to study their living representatives. Studies of paleopathologies associated with decompression syndrome inform us that odontocetes and mysticetes independently evolved specializations for repetitive deep diving. Cross sectional anatomy of ribs from modern and fossil mysticetes indicates that mysticetes started out with hyperdense skeleton and were probably benthic feeding, only to converge on the osteoporosis-like state found among most modern mysticetes and odontocetes. In the end, these studies of fossil cetaceans highlight the fact that many of our assumptions about homologies of anatomy, physiology, and behavior in modern cetaceans may be misled by only studying modern cetaceans, and that interpretations of modern animal biological data that rely on these sorts of assumptions should be reconsidered.

Chapter 6- Advantages of social life relates to protection from predators and foraging efficiency. Disadvantages relate to the need to synchronise one's own activity with that of others. The analysis of which activity is facilitated or detracted in social groups offers an indication of the benefits and costs of solitary or social life, and the pressures individuals suffer in each situation. In this paper we compare the activity budget of adult estuarine dolphins (*Sotalia guianensis*) in two situations, solitary and in a group. Analyses of 197 hours of behavioral sampling (Predominant Activity Sampling) at Pipa Bay located at south Rio Grande do Norte – Brazil support suggestions that protection from predators is facilitated in groups (animals Rest and Travel more in groups), but are obscure regarding advantages of groups for efficiency in foraging. The pressures for social life in estuarine dolphins are discussed in light of the small size of groups observed throughout its area of occurrence.

In: Whales and Dolphins
Editor: Craig A. Murray, pp. 1-46
ISBN 978-1-61668-693-2

Chapter 1

CENOZOIC ENVIRONMENTAL CHANGES AND EVOLUTION OF BALEEN WHALES

Michelangelo Bisconti*
Museo di Storia Naturale del Mediterraneo,
via Roma, Livorno, Italia

ABSTRACT

The earliest baleen-bearing whales (Mammalia, Cetacea, Mysticeti) are represented by fossils found in the Oligocene sediments of New Zealand, Japan, North America, and Europe. Their geographic distribution suggests that the origin of this particular mysticete group occurred in the remaining of the ancient Tethys ocean. During Miocene and Pliocene, baleen-bearing mysticetes dispersed all around the world and gave rise to several evolutionary radiations which are now identified at a taxonomic level in family-rank clades. These are Balaenidae, Balaenopteridae, Cetotheriidae, Eschrichtiidae, and Neobalaenidae. Some taxa are still to be considered incertae sedis and include archaic mysticetes from the Early and Middle Miocene of Japan, United States, Belgium, and Italy. The evolutionary radiations of baleen-bearing mysticetes are related to the development of particular characters linked to highly specialized feeding behaviors. In their turn, these are related to availability of particular kinds of food (i.e., krill, small-sized fish). It is clearly demonstrated that the abundance of krill in particular areas

* Museo di Storia Naturale del Mediterraneo, via Roma 234, 57100, Livorno, Italia. E-mail: zoologia.museo@provincia.livorno.it.

depends upon dynamic characters of oceanic currents, upwelling and other physical features of particular areas. For this reason, it should be possible to find a link between the evolutionary radiations of baleen-bearing whales and the environmental changes occurred in the oceans since their origin in the Late Oligocene time. The goals of this chapter are (a) to review the evolutionary radiations of baleen-bearing mysticetes; and (b) to individuate eventual environmental factors able to influence differential survival of peculiar mysticete forms in particular periods. The final synthesis will provide new ecological hypotheses explaining the increase/decrease of diversity in the group and, eventually, particular pattern of extinction and suggestions for conservation strategies of the living mysticete species.

Keywords: *Baleen whales, Biogeography, Cenozoic, Cetacea, Evolution, Global change, Mysticeti, Phylogeny.*

INTRODUCTION

Baleen whales are large-sized, filter feeding marine mammals belonging to the Order Cetacea. They are included within the infraorder Mysticeti that comprises 10-12 living species grouped within four families: Balaenidae (right and bowhead whales), Eschrichtiidae (gray whale), Neobalaenidae (pygmy right whale), and Balaenopteridae (humpback and rorqual whales) [12, 102]. The few species living today represent only a small amount of the total diversity of the mysticetes if their fossil record is taken into account [49, 120]. Mysticetes are, thus, an example of a taxon characterized by high past diversity and recent decline; this fact makes it necessary to thoroughly consider their fossil record when attempting to reconstruct their phylogenetic, biogeographic, and evolutionary history [127, 128].

Baleen whales include mostly large-sized species and, among them, the largest animal ever lived (the blue whale, *Balaenoptera musculus*). Individuals belonging to the genera *Balaenoptera* (rorquals), *Megaptera* (humpbacks), *Balaena* (bowhead whales), and *Eubalaena* (right whales) often exceed 18-20 m in length and weigh many tons [50, 117]. The smallest living mysticetes are the pygmy right whale (*Caperea marginata*) and the minke whale (*Balaenoptera acutorostrata*) which approach 10-11 m in length [5, 115]. However, the fossil record documents the fact that in the past, most of the mysticete species had lenghts approaching those of the smallest living ones.

Living mysticetes are filter feeders. They use baleen to extract small preys from the water that fulfill their mouths through specialized feeding behaviors. Several authors attempted to describe the feeding behaviors and the internal mechanisms of mysticete feeding but we are still far from having a complete analysis of their tremendous ability to catch hundreds of prey items per time [109]. The fossil record documents that filter-feeding mysticetes were already living in the late Oligocene (around 28-30 Ma) [107, 108] and that they evolved from toothed ancestors that originated in the Eocene [42, 86], just a few million years after the origin of the Order Cetacea [4, 53]. The transition from toothed to baleen-bearing mysticetes has been described in detail in morphological and phylogenetic terms [38].

Living mysticetes are specialized in capturing different fish and crustacean species. Balaenids and neobalaenids filter feed mostly on calanoid copepods in both northern and southern hemispheres but those in the southern hemisphere are able to capture also euphausiid crustaceans (which are collectively known as krill). Southern balaenopterids feed on euphausiids but northern ones are specialized in the capture of different crustacean and fish species, such as *Parathemisto* and *Sardinella* [50]. Gray whales (eschrichtiids) have been described as bottom feeders because they use their rostrum to dredge the seabed to raise small invertebrates in a cloud of mud which then filter [109].

The distribution of the prey species is a key to interpret the biogeographic distribution of living mysticetes. In fact, calanoid copepods, euphausiids, and schooling fish are closely linked to the presence of nutrients which are made available by oceanic currents and upwelling [109]. In their turn, direction and strength of oceanic currents and the intensity of upwelling systems are connected to pattern of temperature and to the geography of ocean basins. In this sense, the biogeographic distribution of modern-day mysticetes is somewhat linked to climatic and geographic patterns because these influence the distributions of their preys.

It has been argumented that the tectonic events that opened the Drake Passage and triggered the Circum Antarctic Current were responsible of the origin of filter-feeding mysticetes in the Eocene-Oligocene time and that the center of origin of the group should be located in the southern Pacific for this reason [43, 44]. It is a matter of fact that the oldest toothed mysticetes are from Antarctica, Japan, California, and Australia (Figures 1 and 2) but the oldest baleen-bearing mysticetes have been recorded from Japan, New Zealand, Austria, Germany, and South Carolina suggesting an early distribution that encompasses the whole Tethys. Is this distribution solely related to the Circum Antarctic Current? Is it possible to explain the evolution of diversity of baleen-bearing mysticetes based upon Cenozoic changes in climate, ocean currents, and geography? In other words, what was the link between baleen whale evolution and Earth global change?

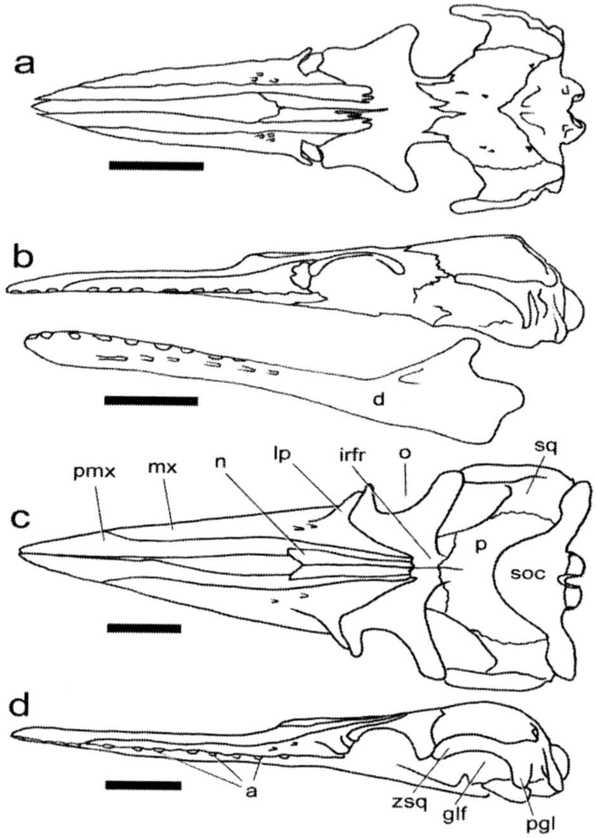

Figure 1. Oligocene toothed mysticetes. *Chonecetus goedertorum* in (a) dorsal and (b) lateral views; *Aetiocetus cotylalveus* in (c) dorsal and (d) lateral views. Scale bars equal 10 cm. Anatomical abbreviations: a, dental alveolus; d, dentary; glf, glenoid fossa of the squamosal; irfr, interorbital region of the frontal; lp, lateral process of the maxilla; mx, maxilla; n, nasal; o, orbit; p, parietal; pgl, postglenoid process of the squamosal; pmx, premaxilla; soc, supraoccipital; sq, squamosal; zps, zygomatic process of the squamosal. (Redrawn from Barnes, L., Kimura, M., Furusawa, H., Sawamura, H. (1994). Classification and distribution of Oligocene Aetiocetidae (Mammalia; Cetacea; Mysticeti) from western North America and Japan. The Island Arc, 3, 392-431 with modifications).

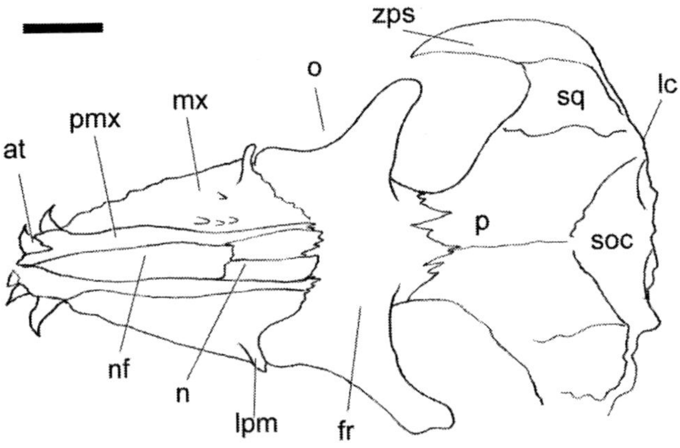

Figure 2. *Janjucetus hunderi* in dorsal view. Scale bar equals 10 cm. Anatomical abbreviations: at, anterior teeth; fr, frontal; lc, lambdoidal crest; lpm, lateral process of the maxilla; mx, maxilla; n, nasal; nf, narial fossa; o, orbit; p, parietal; pmx, premaxilla; soc, supraoccipital; sq, squamosal; zps, zygomatic process of the squamosal. Redrawn from ref. [42] with modifications.

In this chapter, I try to respond to these questions by (1) a description of the principal radiations of baleen whales based upon their fossil record and phylogenetic hypotheses published up to now, (2) a description of the major tectonic events that have influenced the geography of ocean basins in the last 50 million years, (3) an assessment of global climatic changes occurred in the same period, (4) an attempt to reconstruct the nutrient availability over time in different ocean basins, (5) a search for an integrated view of the mysticete evolution and Cenozoic global changes.

Given that many new fossil baleen-bearing mysticete specimens are under description, the results of the present paper have to be intended as provisional. The goal of this study is to provide a basis for further studies in the relationships between the evolution of baleen-bearing mysticetes and global changes, a theme that is thought to influence the survival of modern mysticetes in the next decades.

A Darwinian Approach

Evolution is a biological phenomenon that should be studied and interpreted in biological terms. In this sense, an attempt to explain mysticete evolution by the reconstruction of environmental changes may appear not relevant or formally wrong. It might be clear that, from the methodological standpoint, global changes cannot influence the generation of diversity which is the core of the evolutionary process. The origin of biodiversity depends upon casual genetic mutations but not upon the triggering of an ocean current or the closure of an ocean basin. Environmental changes can influence aspects of the evolutionary history of a group by (1) promoting or depressing speciation through habitat fragmentation, and (2) promoting or depressing the fitness of a species due to variation in the regime of selection (natural selection). For this reason, it must be clear that the impact that global changes may have had on mysticete evolution is limited to the following points:

a. global changes may have been the cause of an increase or a decrease of nutrients leading to population augments or decrements (up to extinction);
b. global changes may have been the cause of habitat fragmentations and isolations of small populations leading to allopatric speciation events or extinctions;
c. global changes may have been the cause of the biogeographic distributions of food items and, indirectly, of mysticetes.

Based on these observations, the only way I have to proceed consists in searching events occurred in the Cenozoic changes in Earth ecology and mysticete evolution in order to assess hypotheses able to explain evolutionary patterns observed in the mysticete fossil record.

Overview of Baleen-Bearing Whale Evolution

The study of the past mysticete diversity started in 18th century and resulted in the discovery of many fossil taxa ranging from species to familty level. A recent attempt to create a new phylogenetic systematics for the baleen-bearing mysticetes resulted in the establishment of additional families [113]. While this attempt is important because of the methodological approach it followed, there are still several new fossil mysticetes under description that could change the present results which are based mainly on the morphology of the earbones. In this sense, the data of that study need to be supplemented with data from the other cranial districts and the postcranial in order to have a more supported phylogenetic view of mysticete systematics. Here, we limit our systematic treatment to those families and subfamilies that are unquestionably well established based on a thorough analysis of their skeletal morphology. However, a discussion of the results mentioned above [113] will be provided in one of the next sections.

Bearing this approach in mind, three other baleen-bearing mysticete families have been formalized, apart from the four living ones: Eomysticetidae, Cetotheriopsidae, and Cetotheriidae. Eomysticetidae was firstly formalized in [107]) to include Late Oligocene baleen-bearing mysticetes from South Carolina and subsequently found also in New Zealand [47] and possibly Japan [93]. Cetotheriopsidae has been established in [52] to include Oligocene baleen-bearing mysticetes from South Carolina and Austria. Cetotheriidae has a more complicated taxonomic history (see reviews in [48, 49, 120] resulting in conflicting interpretations by different authors. Cetotheriidae has been considered a taxonomical basket for many years but now this name is used to include only part of the fossil Oligocene, Miocene and Pliocene mysticetes for which it was established. Cetotheriidae is now used to include *Cetotherium*-like mysticetes such as *Piscobalaena nana, Cetotherium rathkei, Metopocetus durinasus*, *Mixocetus elysius*, *Herpetocetus scaldiensis, H. transatlanticus* and Japanese *Herpetocetus* [27, 65, 67, 92, 97, 125]. In the following text, I will call them cetotheriids. Other mysticetes (*Parietobalaena palmeri, Diorocetus hiatus, Aglaocetus patulus, Isanacetus laticephalus*, etc.: I will call them 'cetotheres') are now placed within Mysticeti incertae sedis [120] or Cetotheriidae incertae sedis [125] because their branching within Cetotheriidae is not supported by all the computer-assisted analyses performed in the recent years and because they are not consistently included within a monophyletic group. The chronological placements of the mysticetes discussed below are from the original papers and from the Paleobiology Database digital archive (mostly compiled by Mark D. Uhen and freely available at http://www.paleodb.org/cgi-bin/bridge.pl).

Morphological and Taxonomic Diversity Changes through Time

The earliest known baleen-bearing whales are from the Late Oligocene Ashley Formation of South Carolina that has been deposited around 28 Ma [107]. They are known as Eomysticetidae (Figure 3). These baleen-bearing whales retains primitive features in the morphology of the neurocranium and earbones resembling basilosaurid archaeocetes very closely. Sanders and Barnes [107] included also an Austrian specimen within Cetotheriopsidae and Fordyce [47] found that many of the New Zealand Oligocene mysticetes were to be assigned to Eomysticetidae too. Sanders and Barnes [108] found also another baleen whale species in the Oligocene of South Carolina, Micromysticetus rothauseni, that they included in Cetotheriidae (Figure 3).

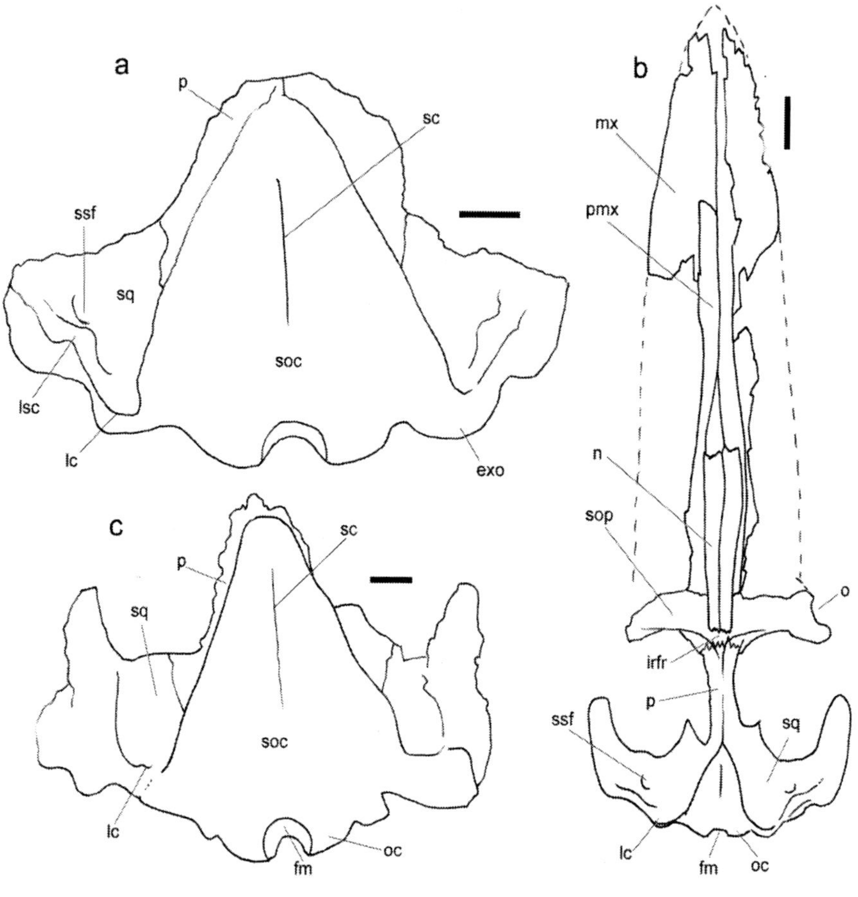

Figure 3. Oligocene baleen-bearing mysticetes. The whole European sample is represented on the left side of the image: (a) *Micromysticetus tobieni* and (b) *Cetotheriopsis lintianus*. (c) *Eomysticetus whitmorei* from South Carolina. Scale bars equal 10 cm. Anatomical abbreviations: exo, exoccipital; fm, foramen magnum; fr, frontal; irfr, interorbital region of the frontal; lc, lambdoidal crest; lsc, lateral squamosal crest; mx, maxilla; n, nasal; o, orbit; oc, occipital condyle; p, parietal; pmx, premaxilla; soc, supraoccipital; sop, supraorbital process of the frontal; sq, squamosal; ssf, secondary squamosal fossa. Redrawn from refs. [107, 108] with modifications.

This species is closely related to a German specimen from the Oligocene previously assigned to Cetotheriopsis tobieni [105]. For the first time, these discoveries allow us to have a view of the early evolution and biogeography of baleen-bearing whales. It is to be noted that the earliest baleen-bearing mysticetes were contemporaneous with the latest toothed mysticetes.

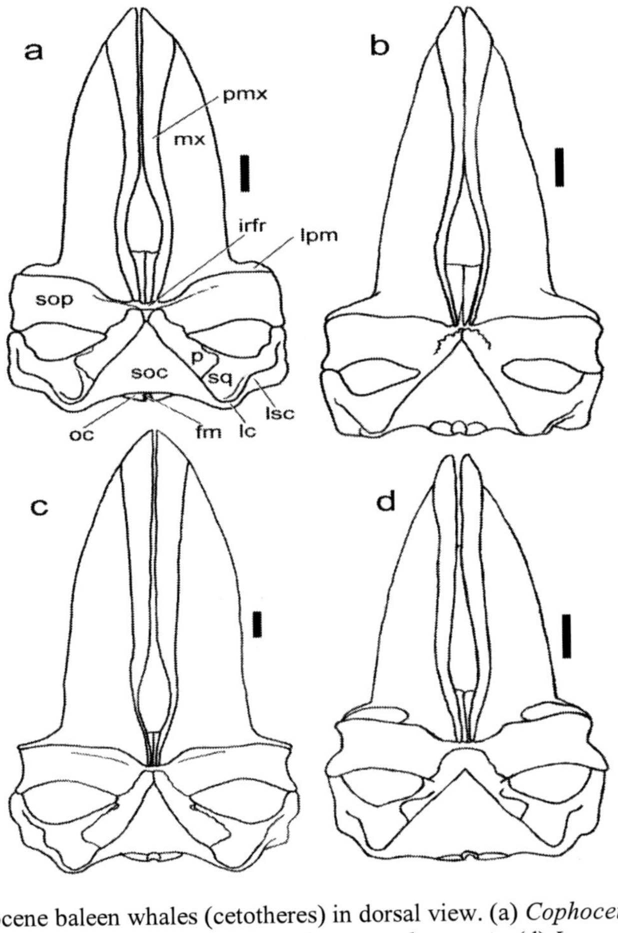

Figure 4. Miocene baleen whales (cetotheres) in dorsal view. (a) *Cophocetus oregonensis*, (b) *Aglaocetus patulus*, (c) *Pelocetus calvertensis*, (d) *Isanacetus laticephalus*. Skulls are drawn at the same cranial width. Scale bars equal 10 cm. Anatomical abbreviations: fm, foramen magnum; irfr, interorbital region of the frontal; lc, lambdoidal crest; lpm, lateral process of the maxilla; lsc, lateral squamosal crest; mx, maxilla; oc, occipital condyle; p, parietal; pmx, premaxilla; soc, supraoccipital; sop, supraorbital process of the frontal; sq, squamosal. Redrawn from refs. [66, 69, 72, 94] with modifications.

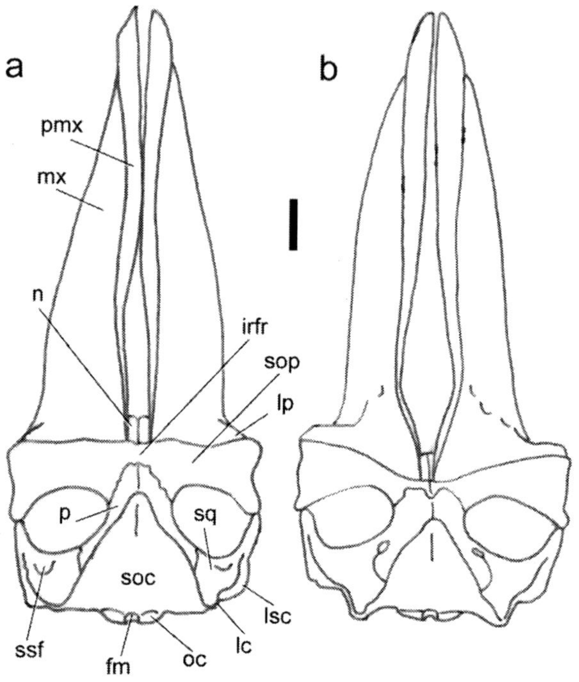

Figure 5. Miocene baleen whales (cetotheres) in dorsal view. (a) *Parietobalaena palmeri*, (b) *Diorocetus hiatus*. Scale bar equals 10 cm. Anatomical abbreviations: fm, foramen magnum; irfr, interorbital region of the frontal; lc, lambdoidal crest; lpm, lateral process of the maxilla; lsc, lateral squamosal crest; mx, maxilla; oc, occipital condyle; p, parietal; pmx, premaxilla; soc, supraoccipital; sop, supraorbital process of the frontal; sq, squamosal. Redrawn from refs. [70, 73] with modifications.

During the Miocene, baleen-bearing mysticetes become highly diversified. The modern-day mysticete families all appeared during the Miocene. Well-dated mysticetes from the earliest Miocene are known from Japan and North and South America. The oldest Miocene mysticetes are representatives of two different radiations. One is represented by one specimen, Morenocetus parvus, from Argentina which is the earliest known balaenid which is from the Burdigalian (around 23 Ma [30]. The others are representatives of a wide radiation of 'cetotheres'. This name identifies a possibly paraphyletic group characterized by transversely large interorbital region, short to absent lateral process of the maxilla, borders of maxilla highly divergent posteriorly giving the rostrum a long transverse diameter, absence of the ascending process of the maxilla, high and wide suprameatal area of the periotic. The 'cetothere' species of the Aquitanian and Burdigalian are Aglaocetus moreni, Cophocetus oregonensis, and Isanacetus laticephalus [64, 72, 94] (Figure 4). For what concerns balaenids, it must be said that Fordyce [46] claimed the presence of a balaenid fossil in the Oligocene of New Zealand but the specimen has not been described up to now.

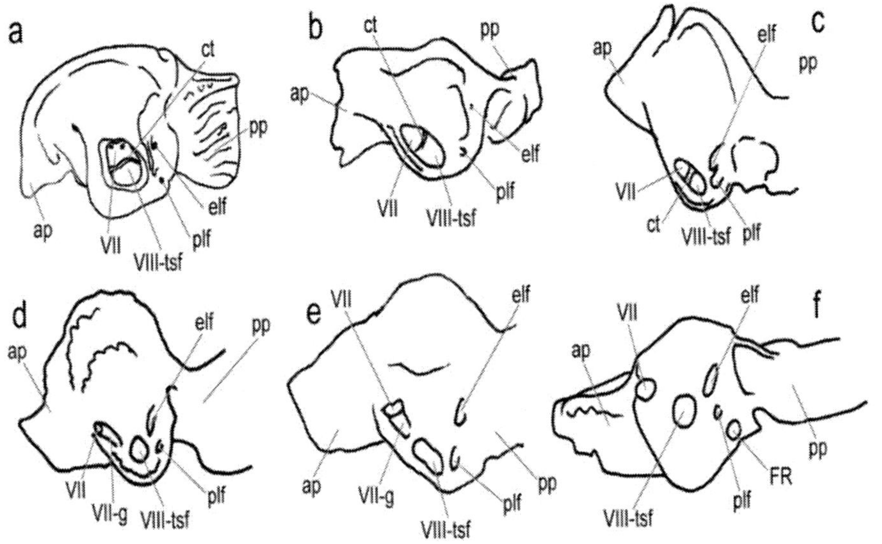

Figure 6. Periotic bones of Balaenomorpha and an archaeocete representative: (a) *Zygorhiza kochii* (Archaeoceti), (b) *Eomysticetus whitmorei* (Eomysticetidae), (c) *Isanacetus laticephalus* ('cetothere'), (d) *Parietobalaena palmeri* ('cetothere'), (e) *Metopocetus durinasus* (Cetotheriidae), (f) *Balaenoptera physalus* newborn (Balaenopteridae). Not to scale. Note that in (a), (b), and (c) the endocranial opening of the facial canal (VII) is separated by a subtle crista transversa from the area where tractus spiralis foraminosus (tsf) and foramen singulare (VIII) open (the internal acoustic meatus). In (d), (e), and (f), VII is widely separated from the internal acoustic meatus and in (d) and (e) a groove runs from the VII opening towards the anterior rim of the internal acoustic meatus. In the newborn fin whale (f) the arrangement of foramina is similar to Cetotheriidae and *P. palmeri* but in adult individuals the arrangement is similar to (a), (b), and (c). However a minority of individuals continue to maintain the Cetotheriidae-like pattern also at adulthood (Bisconti in prep.). Anatomical abbreviations: ap, anterior process; ct, crista transversa; elf, endolymphatic foramen; FR, fenestra rotunda; plf, perilymphatic foramen; pp, posterior process; VII, endocranial opening of the facial canal; VII-g, grove developed from the VII opening to the anterior rim of the internal acoustic meatus; VIII-tsf, foramen singulare-tractus spiralis foraminosus (these opens within the internal acoustic meatus).

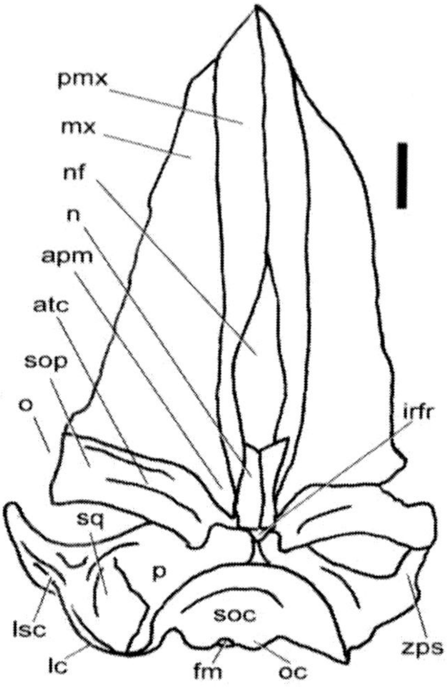

Figure 7. *Titanocetus sammarinensis*, a Miocene baleen whale. Scale bar equals 10 cm. Anatomical abbreviations: apm, ascending process of the maxilla; atc, ascending temporal crest; fm, foramen magnum; irfr, interorbital region of the frontal; lc, lambdoidal crest; lpm, lateral process of the maxilla; lsc, lateral squamosal crest; mx, maxilla; nf, narial fossa; o, orbit; oc, occipital condyle; p, parietal; pmx, premaxilla; soc, supraoccipital; sop, supraorbital process of the frontal; sq, squamosal; zps, zygomatic process of the squamosal. Redrawn from ref. [21] with modifications.

Serravallian mysticetes do not include Balaenidae; there is, in fact, a large gap from the earliest known balaenid Morenocetus parvus and the subsequent latest Miocene and Pliocene balaenids. In the Serravallian and the Langhian, it is observed the persistence of the 'cetotheres' (Pelocetus calvertensis and Aglaocetus patulus) from the North America [66, 69] and the origin of morphological novelties in the anatomy of the earbones and the shape of the rostrum. In Parietobalaena palmeri, a 'cetothere' from the Langhian and Serravallian of the United States [63, 70] (Figure 5), the lateral process of the maxilla are still short to absent and the rostrum lacks a complex interdigitation with the frontal but in the periotic, the endocranial opening of the facial canal becomes separated from the internal acoustic meatus (Figure 6). This morphological novelty is found in the Cetotheriidae and in the developing periotics of Balaenopteridae [17] but is absent in the development of the Balaenid periotic and of 'cetotheres'.

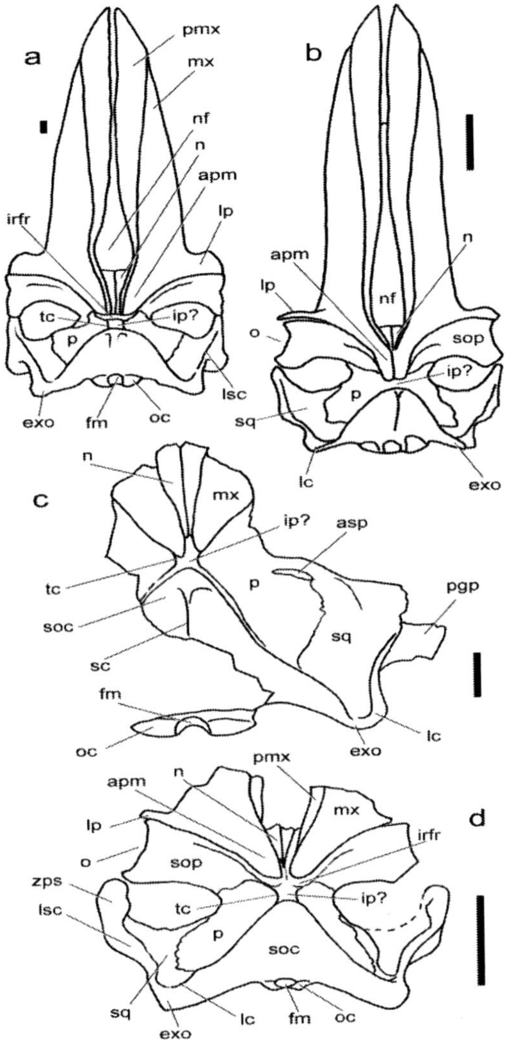

Figure 8. Skulls of Cetotheriidae representatives in dorsal view: (a) *Mixocetus elysius*, (b) *Piscobalaena nana*, (c) *Cetotherium rathkei* (braincase), (d) *Metopocetus durinasus* (braincase). Scale bars equal 10 cm. Anatomical abbreviations: apm, ascending process of the maxilla; asc, alisphenoid; atc, ascending temporal crest; exo, exoccipital; fm, foramen magnum; ?ip, possible interparietal; irfr, interorbital region of the frontal; lc, lambdoidal crest; lpm, lateral process of the maxilla; lsc, lateral squamosal crest; mx, maxilla; nf, narial fossa; o, orbit; oc, occipital condyle; p, parietal; pmx, premaxilla; sc, sagittal crest; soc, supraoccipital; sop, supraorbital process of the frontal; sq, squamosal; tc, temporal crest; zps, zygomatic process of the squamosal. Redrawn from refs. [27, 65, 67, 97] with modifications.

In Titanocetus sammarinensis, which is the only Serravallian 'cetothere' from the Mediterranean basin [21], the lateral process of the maxilla is substantially absent and the rostrum is wide at its base but the posteromedial corner of the maxilla is highly elongated forming a distinct ascending process (Figure 7). This feature is observed in the Balaenopteridae, Cetotheriidae, and Eschrichtiidae [27, 118] and should be considered as a key to interpret the origin of the rostrum-neurocranium interface in these families. Unfortunately, the earbones of Titanocetus are not known.

In the Langhian and Serravallian 'cetothere' Diorocetus hiatus from the United States [68], the periotic bears the same modification as described in Parietobalaena palmeri (Figure 6). In addition, in this species the borders of the maxilla are much more straight and parallel to the longitudinal axis of the skull making it the rostrum much more narrow at its base (Figure 5). This feature is observed in the Cetotheriidae and some Balaenopteridae.

Parietobalaena, Titanocetus, and Diorocetus represent transitional taxa that are of much help in understanding the origin of the non-Balaenid modern-day mysticete families and the extinct Cetotheriidae. The basal 'cetotheres' went extinct by the end of the Serravallian (around 11.6 Ma) being substituted by the Cetotheriidae, Balaenopteridae, and Eschrichtiidae all of which appeared in a range of a few million years from the Serravallian to the Tortonian.

Cetotheriidae is a family of baleen-bearing mysticetes characterized by long ascending process of the maxilla whose lateral border is directed anteriorly and laterally giving the rostrum-frontal articulation a 'V' shape (Figure 8); the lateral process of the maxilla is short and anteriorly excavated to form an anterior concavity (antorbital notch); the lateral border of the maxilla is straight and mostly parallel to the longitudinal axis of the skull; the supraoccipital has a rounded anterior border with raised sites for the attachment of the neck musculature. Cetotherium rathkei, Piscobalaena nana, Nannocetus eremus, Mixocetus elysius, Metopocetus durinasus, Herpetocetus scaldiensis, H. transatlanticus, and Herpetocetus sendaicus are typical representatives of this family. Cetotherium rathkei is probably the oldest typical Cetotheriidae being from the Serravallian of the Russian Federation [28, 97]. The periotic of Piscobalaena nana, Metopocetus durinasus, Herpetocetus scaldiensis and H. transatlanticus are small-sized and show that endocranial opening of the facial canal and internal acoustic meatus are separate [27, 67, 125] (Figure 6). Many other putative cetotheriid taxa are known from the southern border of the North Sea as they are conserved in the collections of Brusseles and of Dutch museums but most of them are in need of a critical taxonomical revision after more than one century since their first descriptions [37].

Characters of the dentary suggest that at least some Cetotheriidae were able to perform a different kind of filter-feeding based upon a protrusion of the mandible [125]. This hypothesis is based upon the observation of the long and posteriorly-protruded angular process of the dentary of Herpetocetus. Also Piscobalaena nana shows an elongated angular process but at a lower degree of development [27]. A slight ventral and posterior protrusion of the angular process has been described also in early-diverging 'cetotheres' such as Diorocetus hiatus and an unidentified specimen from Japan [71].

Herpetocetus and Piscobalaena crossed the boundary between Miocene and Pliocene. Finds from California and Belgium show that it survived also in the Pliocene [36, 121, 125]. Presently, they are the only cetotheriid taxa which are found in Pliocene sediments.

In the early 'cetotheres' the parietals meet along the midline of the skull forming a sagittal crest in some species (i.e., Diorocetus hiatus, Parietobalaena palmeri). In the Cetotheriidae, the dorsal surface of the skull that is located anterior to the supraoccipital show a laterally concave relief which is a bilateral formation. A nearly flat surface is found between the laterally concave relieves which are found in the place of the sagittal crest observed in Diorocetus and other 'cetotheres'. This same morphology is observed in the living gray whale, Eschrichtius robustus [82], the only living species belonging to Eschrichtiidae [130]. Based on comparative analysis, it is likely that the dorsal surface located between the laterally concave relieves is the interparietal in both Eschrichtiidae and Cetotheriidae. This character is found also in Titanocetus sammarinensis [21] suggesting a close connection between this Mediterranean taxon with Cetotheriidae and Eschrichtiidae. A split of the sagittal crest into two externally concave relieves separated by a more or less wide interposing surface in part occupied by the interparietal is observed also in Balaenopteridae [122]. Balaenopteridae, Eschrichtiidae, and Cetotheriidae share this particular organization of the vertex of their skulls. Moreover, they share also the organization of endocranial foramens in the periotics (at least during ontogeny in balaenopterids), and the presence of long and definite scending processes of the maxilla. Cetotheriidae and Eschrichtiidae share also the shape of the antorbital notch and the posteriorly protruded exoccipital whose posterolateral border is very close to the postglenoid process of the squamosal in ventral view.

The oldest known Eschrichtiidae is Archaeschrichtius ruggeroi from the Tortonian of southern Italy. This taxon is represented by a single dentary with very peculiar morphology of the coronoid region that shows the presence of a coronoid process, a satellite process of the coronoid process, and a well developed postcoronoid crest and fossa [26]. The latter two features are typically found in Parietobalaena palmeri, Diorocetus hiatus, Titanocetus sammarinensis, Piscobalaena nana, in the early balaenopterid Megaptera hubachi [34], and in another Tortonian balaenopterid not yet published (pers. obs.). The postcoronoid crest and fossa of the dentary are lost in the Pliocene eschrichtiid Eschrichtioides gastaldii from the Pliocene of northern Italy [23] (Figure 9). In this species and in living Eschrichtiidae strongly reduced coronoid and satellite processes are found [26].

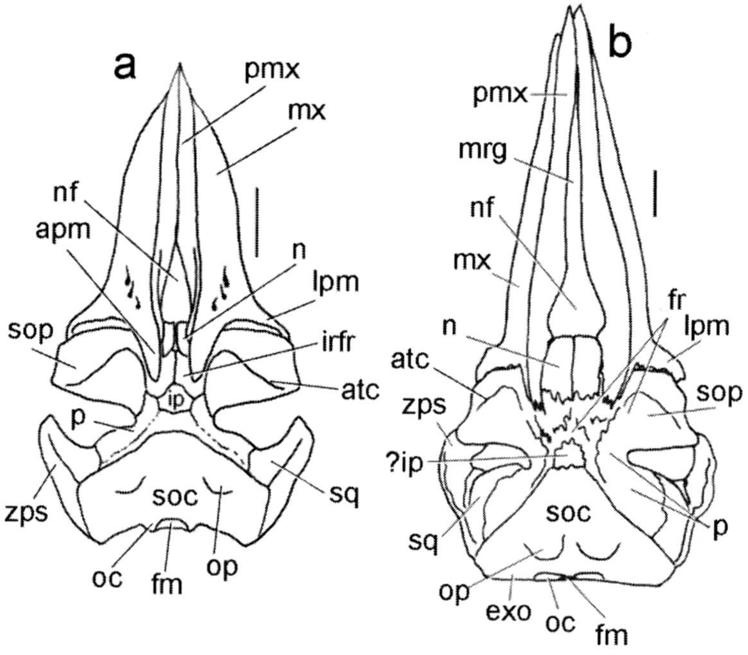

Figure 9. Skulls of Eschrichtiidae representatives in dorsal view: (a) *Eschrichtioides gastaldii*, (b) *Eschrichtius robustus*. Scale bars equal 10 cm. The skulls are represented at the same transversal width. Anatomical abbreviations: apm, ascending process of the maxilla; atc, ascending temporal crest; exo, exoccipital; fm, foramen magnum; ip-?ip, interparietal-possible interparietal; irfr, interorbital region of the frontal; lc, lambdoidal crest; lpm, lateral process of the maxilla; lsc, lateral squamosal crest; mrg, mesorostral groove; mx, maxilla; nf, narial fossa; o, orbit; oc, occipital condyle; ot, occipital protrusion; p, parietal; pmx, premaxilla; soc, supraoccipital; sop, supraorbital process of the frontal; sq, squamosal; zps, zygomatic process of the squamosal. Living gray whale redrawn from ref. [18] with modifications.

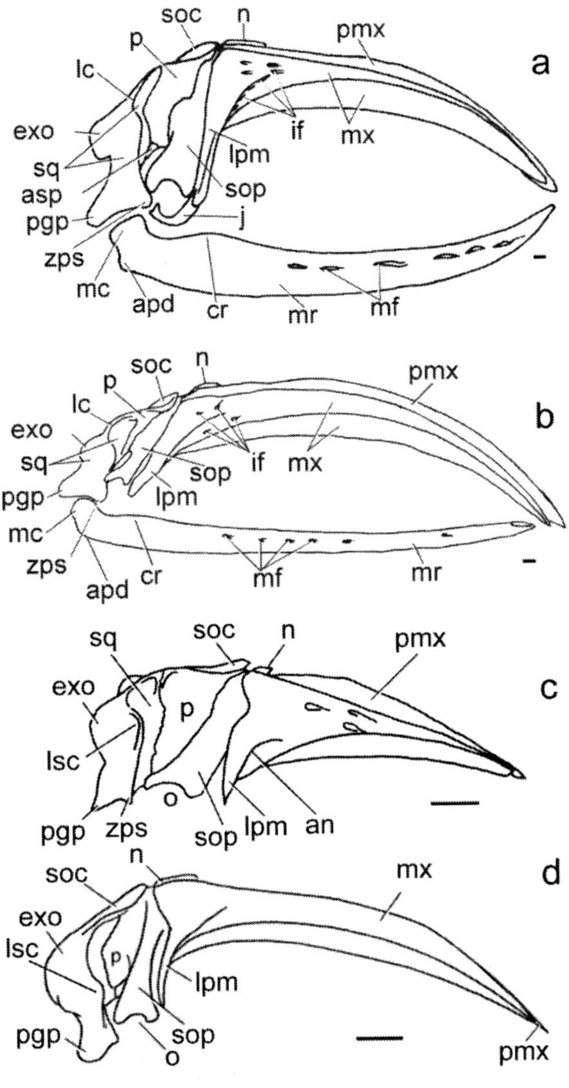

Figure 10. The skull of the minke whale (*Balaenoptera acutorostrata*) as to represent Balaenopteridae in (a) dorsal and (b) lateral views. Scale bars equal 10 cm. Anatomical abbreviations: apm, ascending process of the maxilla; atc, ascending temporal crest; fm, foramen magnum; irfr, interorbital region of the frontal; lc, lambdoidal crest; lpm, lateral process of the maxilla; lsc, lateral squamosal crest; mx, maxilla; nf, narial fossa; o, orbit; oc, occipital condyle; p, parietal; pmx, premaxilla; soc, supraoccipital; sop, supraorbital process of the frontal; sq, squamosal; zps, zygomatic process of the squamosal.

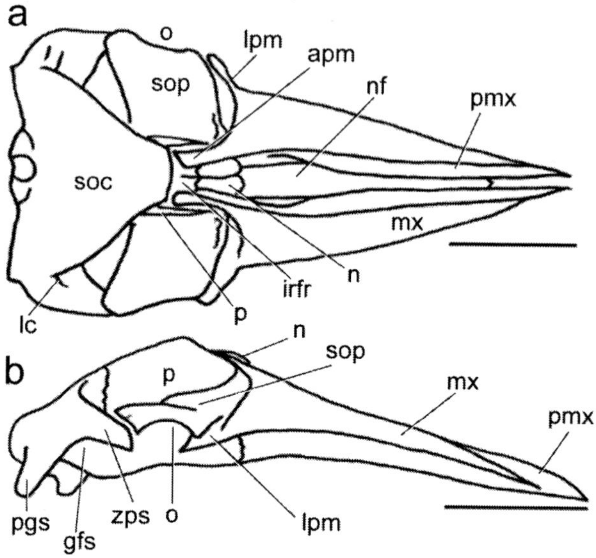

Figure 11. Skulls of Balaenidae representatives in lateral view: (a) *Eubalaena glacialis* (right whale), (b) *Balaena mysticetus* (bowhead whale), (c) *Balaenella brachyrhynus* (reconstruction), (d) *Balaenula astensis* (reconstruction). Scale bars equal 10 cm. Anatomical abbreviations: an, antorbital notch; apd, angular process of dentary; cr, coronoid relief of dentary; exo, exoccipital; if, infraorbital foramens; j, jugal; lc, lambdoidal crest; lpm, lateral process of the maxilla; lsc, lateral squamosal crest; mc, mandibular condyle; mf, mental foramina; mr, mandibular ramus; mx, maxilla; nf, narial fossa; o, orbit; oc, occipital condyle; p, parietal; pgp, postglenoid process of the squamosal; pmx, premaxilla; soc, supraoccipital; sop, supraorbital process of the frontal; sq, squamosal; zps, zygomatic process of the squamosal. Redrawn from [18, 19] with modifications.

Balaenopteridae are known since the late Serravallian and Tortonian (around 12-10 Ma). The oldest balaenopterid taxa include the Tortonian Megaptera miocaena and the Messinian Parabalaenoptera baulinensis from California [62, 132]. Megaptera hubachi, Archaebalaenoptera castriarquati, Protororqualus cuvieri and the Japanese balaenopterid described in [92] are from the Pliocene and represent archaic balaenopterid taxa (Figure 10).

Apart from rare and fragmentary balaenid finds in the Late Miocene of South America and Japan and the Early Miocene Morenocetus parvus, only Pliocene records document the diversity of this family in the past. Recent examinations of the phylogenetic history and the taxonomy of the fossil record of Balaenidae revealed that several genera were living during the Pliocene [16, 18, 19, 81] (Figure 11). Two radiations are found in the Balaenidae based on phylogenetic analysis: one including Balaena and Balaenella and the other including Balaenula and Eubalaena. It is likely that the situation will be complicated when other taxa are included within the analysis (e.g., Balaenotus) and a redescription of Morenocetus is really needed in order to clarify its position. Morenocetus should be considered as more closely related to Balaena and Balaenella than Eubalaena and Balaenula but the support to this conclusion is scanty and additional information are necessary [19].

Up to now, none is known about the fossil record of Neobalaenidae [48, 49], thus the past history of this family is totally unknown. If one accepts its position as the sister group of Balaenidae then it might be originated in the Late Oligocene (before the divergence of the Early Miocene Morenocetus parvus); alternatively, if one accepts its position as the sister group of Balaenopteridae + Eschrichtiidae then it is difficult to hypothesize its divergence date because the sequence of sister groups leading to the modern rorquals and gray whales is still highly debated (see next section).

Hypotheses of Phylogeny

In the last quarter of century several authors attempted to reconstruct the phylogeny of mysticetes based on the fossil record. Molecular studies were also used to this scope suggesting different relationships and stimulating new and more detailed morphological analyses. The new wave of mysticete studies was certainly triggered by a ponderous work on the phylogenetic relationships of the bowhead whale, *Balaena mysticetus*, published in the early 90s [81]. In that work, a phylogenetic tree was published in which balaenids and neobalaenids are sister groups, eschrichtiids are more closely related to balaenids and neobalaenids than to balaenopterids that are the sister group of cetotheriids *s.l.* This result was challenged by new morphological analyses of the mysticete earbones performed subsequently [51]. The new work supported a closer relationship of balaenids and archaic mysticetes (cetotheriids and 'cetotheres') to the exclusion of balaenopterids. Another study published just a few years later [16] substantially supported the results of McLeod and colleagues [81].

It is a matter of fact that a real advancement was stimulated by the cladistic analysis accompanying the description of *Isanacetus laticephalus* [72]. In that work, all the characters traditionally used for describing the evolutionary history of mysticetes were employed in a computer-assisted phylogenetic study. For the first time in the attempts to reconstruct the phylogeny of baleen-bearing mysticetes, balaenids were placed at the base of baleen-bearing whales (at that time Eomysticetidae was not yet known) and Eschrichtiidae and Balaenopteridae were sister groups. Kimura and Ozawa [72] recognized four groups of archaic baleen-bearing mysticetes: an early diverging clade corresponding to Cetotheriidae, a clade formed by *Diorocetus* and *Parietobalaena*, and a paraphyletic group including *Aglaocetus patulus* and *Cophocetus oregonensis*, and *Pelocetus calvertensis* and *Isanacetus laticephalus*.

In the study of the Oligocene *Eomysticetus* [108], the phylogenetic analysis resembled very closely that published in [81] in that Eschrichtiidae is sister group of a clade including Neobalaenidae and Balaenidae. Ceotheriidae is sister group to Balaenopteridae being Cetotheriidae here intended as a basket including both Cetotheriidae *sensu stricto* (*Cetotherium rathkei*, *Mixocetus elysius*, *Metopocetus durinasus*, *Herpetocetus scaldiensis*, *H. sendaicus, Piscobalaena nana*, *Nannocetus eremus*) and the more archaic 'cetotheres' (*Diorocetus*, *Parietobalaena*, *Aglaocetus*, etc). A variation on this result was provided in ref. [52] where it is found that Balaenidae and Neobalaenidae were sister groups, Balaenopteridae and Eschrichtiidae were sister groups, *Diorocetus*, *Pelocetus*, and *Parabalaenoptera* were stem baleen-bearing mysticetes. Exactly the same pattern was found by Fitzgerald [42] that, together with Deméré et al. [35], added a large sample of toothed mysticetes to the analysis.

In 2005 two papers appeared with simultaneous analyses of living and fossil mysticetes. In the paper of Bisconti [19] the focus was on Balaenidae and the results have been summarized in the previous paragraph. It must be noted that in this study Balaenidae and Neobalaenidae (forming the superfamily Balaenoidea) diverged very early and only a few other mysticete taxa were used. In the paper of Deméré et al. [37] a more comprehensive taxonomic sample was used. The cladistic analysis resulted in the discovery that Balaenopteridae and Eschrichtiidae are sister groups, Balaenoidea (including Balaenidae and Neobalaenidae) is sister group of Balaenopteridae + Eschrichtiidae, Cetotheriidae *s.s.* diverged just after Eomysticetidae, and the other 'cetotheres' are considered sedis mutabilis. Only *Diorocetus hiatus* is the sister group of the clade formed by Balaenoidea + Eschrichtiidae + Balaenopteridae.

The 2006 paper on *Piscobalaena nana* [27] included a phylogenetic analysis of baleen-bearing mysticetes. The published cladogram recognized an early diverging Cetotheriidae *s.s.* which branched off the cladogram after Eomysticetidae. Balaenidae and Neobalaenidae are sister groups, Eschrichtiidae is the sister group of Balaenoidea, Balaenopteridae is the sister group of Eschrichtiidae + Eschrichtiidae, and the other 'cetotheres' are placed between Cetotheriidae and Balaenopteridae. In particular, *Aglaocetus patulus* is the sister group of Balaenopteridae + Eschrichtiidae + Balaenoidea; *Diorocetus hiatus* is the sister group of the latter plus all the other 'cetotheres'.

In 2007 two other papers appeared with new analyses of mysticete phylogeny. In the first [113] the cladistic results were used to establish a new taxonomy of mysticete families. In this paper, for the first time, it is cladistically supported a close relationship between Eschrichtiidae and Cetotheriidae. Such a relationship was suggested by earlier studies [2]. Moreover, in that work, Balaenidae and Neobalaenidae branch off the cladogram after a clade including *Eomysticetus* and *'Mauicetus' lophocephalus*; Balaenopteridae are the sister group of a clade including *Diorocetus hiatus* and other two 'cetothere' taxa; other 'cetotheres' are placed in different quite well resolved clades. Steeman named the superfamily Cetotherioidea to include Cetotheriidae and Eschrichtiidae, and included Balaenopteridae, Pelocetidae, Diorocetidae, and Aglaocetidae within the superfamily Balaenopteroidea.

The other paper appeared in 2007 focused on the phylogeny of Balaenopteridae [22]. In that paper, Bisconti reinforced the relationship between Eschrichtiidae and Cetotheriidae, the basal position of Balaenoidea and placed 'cetotheres' as sedis mutabilis at the base of a clade including Cetotheriidae, Eschrichtiidae, and Balaenopteridae being Balaenopteridae the sister group of Eschrichtiidae + Cetotheriidae [22, 23].

Finally, in 2008 a supermatrix study [38] appeared showing for the first time that 'cetotheres' and Cetotheriidae form a monophyletic and early diverging clade. Balaenidae is the sister group of a clade including Neobalaenidae, and an unresolved node including Balaenopteridae and Eschrichtiidae.

In summary, there are different opinions regarding the phylogeny of baleen-bearing mysticetes. The main points of divergence between different authors are the following: (1) the relationships of Balaenidae; (2) the monophyly of Balaenoidea; (3) the relationships of Cetotheriidae; (4) the relationships of Eschrichtiidae. It is to be hoped that the description of other fossil taxa and more detailed morphological analyses will help.

Molecular studies did not help in resolving these problems but added new and different views. Apart from some early works [84, 85] whose results have been largely dismissed [83, 51], the sequence comparisons of mtDNA and an increasing number of nuclear genes suggest the following results: (1) Neobalaenidae is the sister group of Balaenopteridae and Eschrichtiidae, (2)

Eschrichtiidae is included within Balaenopteridae, (3) the genus *Balaenoptera* is not monophyletic [3, 80].

At least points (2) and (3) cannot be defended on a morphological basis but point (1) is supported at least by one supermatrix-based study including both molecular and morphological data [38]. This raises the question of the phylogenetic relationships of Neobalaenidae that must be solved not only in molecular and combined studies but also in morphological ones. Otherwise, the suspicion remains that thousands of molecular data can influence the combined result minimizing the phylogenetic signal contained in the morphological data.

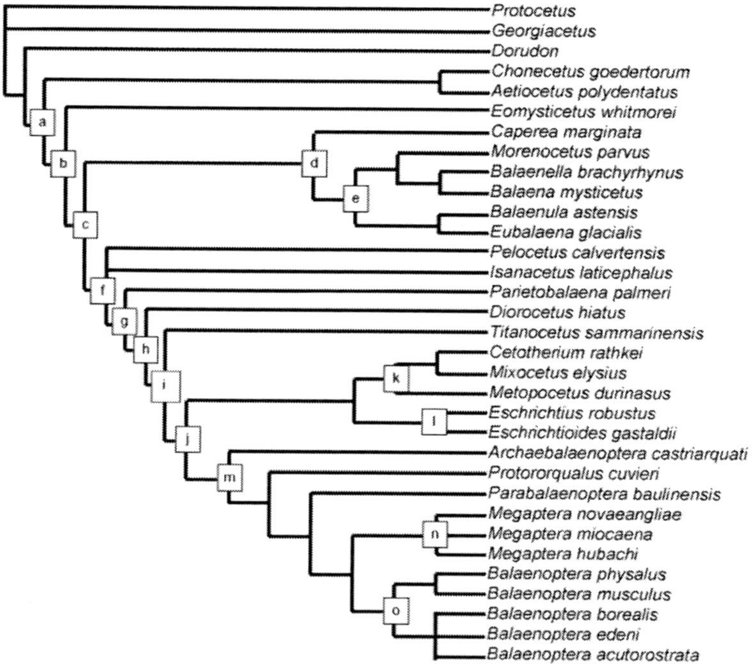

Figure 12. Hypothesis of phylogenetic relationships of living and fossil baleen whales (Balaenomorpha) based on morphology (from refs. [19, 23] with modifications). Nodes: a, Mysticeti; b, Balaenomorpha (baleen whales); c, unnamed clade; d, Balaenoidea; e, Balaenidae; f, unnamed clade; g, unnamed clade; h, unnamed clade; i, unnamed clade; j, unnamed clade; k, Cetotheriidae; l, Eschrichtiidae; m, Balaenopteridae; n, Megapterinae; o, Balaenopterinae.

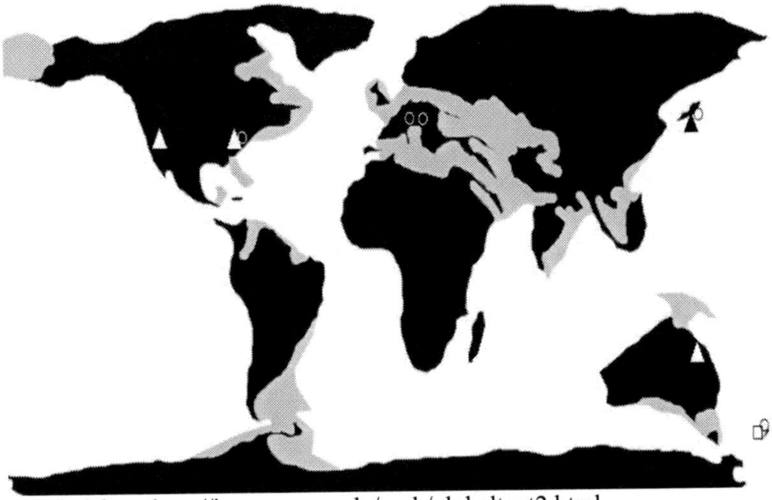

Map modified from http://jan.ucc.nau.edu/~rcb/globaltext2.html.

Figure 13. Distribution of mysticete finds in a schematic map of the Oligocene (35 Ma). Squares represents balaenids, circles represents eomysticetids, triangles represent toothed mysticetes.

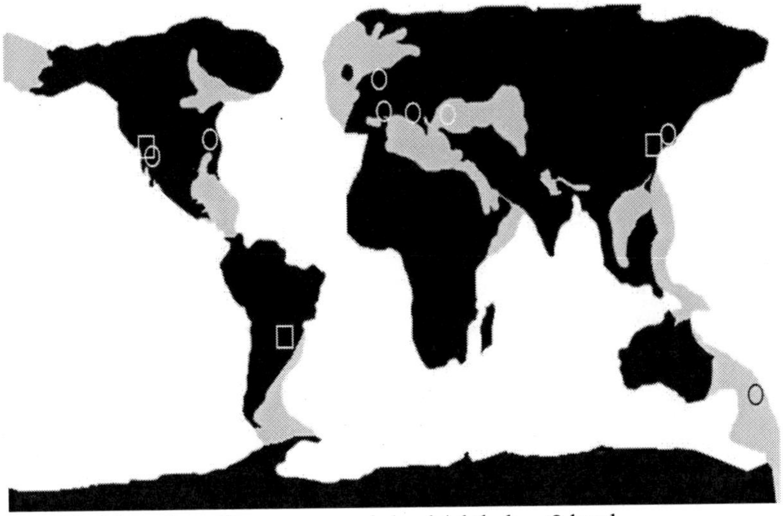

Map modified from http://jan.ucc.nau.edu/~rcb/globaltext2.html.

Figure 14. Distribution of mysticete finds in a schematic map of the Miocene (20 Ma). Squares represent balaenids, circles represent other baleen whale families.

Bisconti [22, 23] calculated the Stratigraphic Consistency Index [60] using many published cladograms to test the congruence between the branching patterns and the stratigraphic occurrence of the taxa. He found that the branching pattern that best fits with the order of stratigraphic occurrence of fossil species is published in refs. [22, 23]. This means that in this cladogram the number of ghost lineages to be taken in interpreting the evolution of the group is smaller. This is the only reason because I will use this cladogram in the next sections (Figure 12). Since it was not possible to find a consensus on the phylogeny of the Mysticeti, in this study I prefer to use a cladogram that, in addition to presenting a hypothesis of phylogenetic relationships, also has the advantage of showing the best agreement with the order of appearance of the taxa.

Paleobiogeographic Distribution

The earliest-diverging baleen-bearing mysticetes are the Eomysticetidae which include specimens from South Carolina, New Zealand, Japan, Austria, and Germany [52, 107, 108, 120]. The paleobiogeographic distribution of this family encompasses most of the Oligocene Tethys suggesting a quick diffusion of the group from a center of origin located probably in the southern portion of the Pacific Ocean [45] (Figure 13). This paleobiogeographic hypothesis is supported by the observation that almost all the toothed mysticetes (Aetiocetidae, Llanocetidae, Mammalodontidae, Janjucetidae) are found in the Pacific Ocean and the oldest forms (Llanocetidae and Janjucetidae) are from Antarctica and Australia [42, 86]. However, at least another family of toothed mysticetes were present in the North Atlantic in the Late Oligocene [7]. Based upon current knowledge, both toothed mysticetes and Eomysticetidae went extinct at the end of the Oligocene when 'cetotheres' started to invade most of the seas.

'Cetotheres' are known from the Miocene of Japan, New Zealand, East and West Coasts of United States, Argentina, Peru, Italy, Portugal, Belgium, Holland, England, Georgia (Caucasus), and Russian Federation (Crimean Peninsula) (Figure 14). Despite this wide family-to-superfamily rank distribution, when one attempts to analyse the distribution of individual taxa the results is very different. In fact, 'cetothere' species show often a limited distribution and occur in single deposits. Only two Miocene genera show a broader range of distribution, *Parietobalaena* and *Diorocetus*, occurring in US East Coast and Japan [68, 70, 73, 74] and *Aglaocetus* occurring in US and

Argentina [69, 64]. The result is that the mysticete faunas from the Miocene are formed mainly by endemic taxa.

In the Pliocene, also, a few genera are found in different basins: *Balaenula* (Italy, Belgium, US East and West Coast, Japan), *Balaena* (Italy, Belgium, Japan, US East Coast), *Herpetocetus* (Belgium, US East and West Coast, Japan). In this sense, also the Pliocene faunas were mainly formed by endemic taxa and the descripions of new balaenopterid and balaenid genera from Italy, Belgium and US reinforce this observation. In particular, it also appears that the Mediterranean fauna was largely isolated from the other basins during the Pliocene because its balaenopterid fauna is characterized by basal balaenopterids which are more primitive than those inhabiting the US West Coast and South America [25]. In this sense, the Mediterranean played an important role in the preservation of ancient biodiversity in the Pliocene.

Data about the Quaternary evolution of mysticetes are very poor. Only a few attempts have been made to review the Pleistocene and Holocene records of modern species but these are of particular interests because help in reconstructing the early history of the living species. Reviews on the Quaternary records of Eschrichtiidae revealed that the gray whale was distributed in the North Atlantic up to historical times [6, 77]. Its extinction is related probably to massive hunting performed by humans. Today, the gray whale is distributed only along the northern margin of the Pacific Ocean.

None is known about the origin and paleobiogeographic history of the modern balaenopterid species. A few Neogene records of *Megaptera* have been reported in isolated studies from Japan (but the taxonomy was uncertain) [91] and Australia [41] but the origin of the modern *Balaenoptera* species is virtually unknown. A Pliocene balaenopterid with the skull resembling that of the living forms is known from Japan (Maesawa-Cho) [92]; this suggests that the center of origin of living *Balaenoptera* species was located possibly in the Pacific Ocean. However, this observation has to be carefully tested by detailed morphological and phylogenetic analyses of the specimen. There are no data regarding the origin of the antitropical distribution of modern balaenopterid mysticetes. Today, it is known that these whales are distributed in northern and southern stocks which are connected by a small gene flow [59]. These stocks move yearly from precise feeding areas to precise reproductive areas and each stock has its own culturally-inherited communication signals and behaviors [24, 90]. This suggests a sort of tendence towards a genetic insulation of each stock that could be the prelude of new speciation events. In other words, the fragmentation of the distribution of modern balaenopterids in precise and isolated areas inhabited by different stocks can be the prelude of future allopatric speciation events leading to the origins of new species with limited ranges.

There are very limited data about the historical distributions of balaenid species. Quaternary records have been reviewed by MacLeod et al. [81] but the centers of origin of the living species are not known. Genetic studies have suggested that changes in the range of the right whale species inhabiting the North Pacific (*Eubalaena japonica*) could have influenced the ability of this species to colonize southern oceans from northern areas leading to the origin of the southern right whale (*E. australis*) [104]. Such distribution changes could be due to changes in climatic regime during the last glacial periods [104].

Our understanding of the paleobiogeography of the modern-day mysticete species will be highly influenced by our ability to put the evolutionary history of the living species in a broader context which has to include the fossil record. In this sense, each new discovery, even fragmentary, of Quaternary mysticetes is necessary and should be published. In fact, the most recent history of mysticete whales is probably less known than the early divergence events dating about 28 Ma.

Connecting Cenozoic Global Changes and Baleen Whale Evolution

The origin of baleen-bearing whales (Balaenomorpha) [52] has been put into relation with the triggering of the Antarctic Current that changed the distribution and availability of nutrients able to support large populations of schooling fish and crustaceans [43, 44]. This hypothesis supports the view that southern Pacific was the center of origin of the mysticetes; an additional support to this view is provided by the discovery of several toothed mysticetes from Oligocene deposits scattered around the Pacific ocean.

Very few attempts have been made to explain the Miocene and Pliocene diversity and paleobiogeographic distribution of baleen-bearing whales and to understand their extinction patterns [18, 19]. In the following text, an integrated view of mysticete phylogeny and distribution and Earth evolution in the Cenozoic is provided in the search for reciprocal influences able to explain the evolutionary history of this marine mammal group.

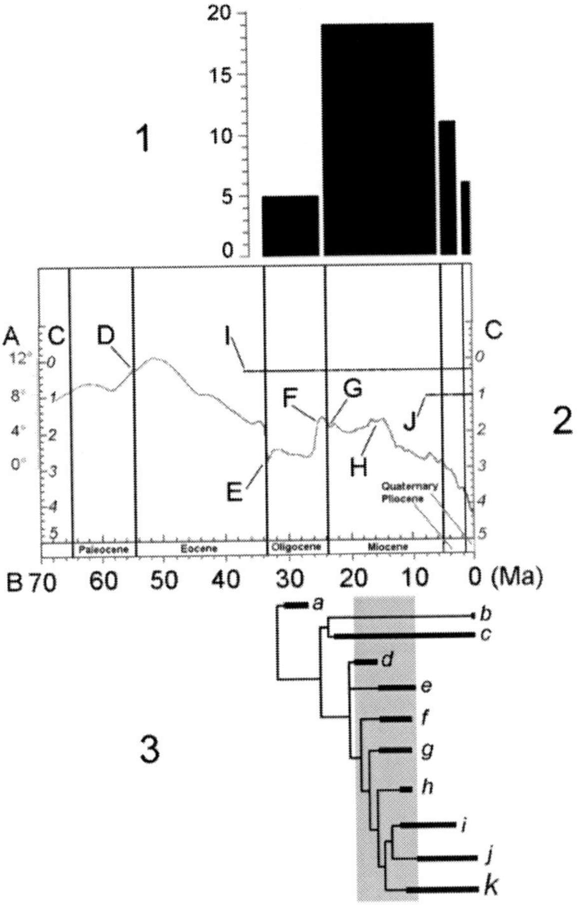

Figure 15. Relationship between phylogeny, diversity change and temperature modifications during the Cenozoic. 1, mysticete diversity over time plotted on the prolongation of period limits as in 2; numbers on the left are number of baleen whale genera (see text for details). 2, principal climatic events occurred in the Cenozoic based on $\delta^{18}O$ (A, numbers represent difference in temperatura from current mean temperature in °C; B, million years ago (Ma); C, $\delta^{18}O$ values (modified from [131]); D, Late Paleocene and Early Eocene thermal optima; E, *Oi-1* glaciation; F, Late Oligocene warming; G, *Mi-1* glaciation; H, mid-Miocene climatic optimum; I, Antarctic ice-sheets; J, northern hemisphere ice-sheets. 3, simplified mysticete phylogeny (only baleen-bearing mysticetes are represented); gray area shows the period of maximum origination rate occurred between 20 and 10 Ma (a, Eomysticetidae; b, Neobalaenidae; c, Balaenidae; e, *Isanacetus*; e, *Pelocetus*; f, *Parietobalaena palmeri*; g, *Diorocetus hiatus*; h, *Titanocetus sammarinensis*; i, Cetotheriidae; j, Eschrichtiidae; k, Balaenopteridae); black rectangles represent the age of known fossil records (see text for details).

Oligocene: Early Radiation

The initial diffusion of Oligocene Balaenomorpha (baleen-bearing mysticetes) is recorded by several finds in New Zealand, Austria, Germany, South Carolina and Japan (Figure 13). Oligocene Balaenomorpha include only Eomysticetidae and Cetotheriidae (*sensu* [107, 108]) but the cetotheriids represented in the record (*Micromysticetus*) are quite different from the Miocene 'cetotheres' and Cetotheriidae *s.s.* It is still unclear if *Micromysticetus* represents an early-diverging cetotheriid or if it is more closely related to Eomysticetidae.

Oligocene Balaenomorpha exhibited low diversity but attained a world-wide distribution in a relatively short time span. Balaenomorpha should be originated around 30-34 Ma in the lower Oligocene [42]. Sanders and Barnes found eomysticetids and *Micromysticetus* in 30 Ma deposits in South Carolina [107, 108]; Fordyce reported eomysticetid finds in New Zealand from 24-28 Ma deposits in the upper Oligocene [47]; the European and Japanese records of eomysticetids and *Micromysticetus* are from the late Oligocene [105, 107]. All these finds suggest that in the late Oligocene, eomysticetids and *Micromysticetus* were well established all over the Tethys and that the origin of Balaenomorpha must be somewhat older.

It is noticeable that the Oligocene baleen-bearing whales exhibited so low diversity: only two genera are currently recognized with a handful of species, some of which are still to be described. The distribution of Oligocene Balaenomorpha in such a wide areal points to the possibility that new species and genera may be found and described.

It is still unclear if northern and southern hemisphere populations were segregated at this time [44] because while eomysticetids are known from northern and southern areas, their taxonomy is not still completely understood; publication of a revision of the New Zealand eomysticetids recently cited [47] would be of great help.

It has been argued that the total separation of Australia from Antarctica occurred in the Oligocene [31] worked as a trigger for mysticete evolution. In fact, the megatectonic events affecting this part of the planet were responsible of the initiation of the Circum-Antarctic current which moved nutrients northwards making them available for exploitation by phytoplankton and zooplankton. The abundance of food resulting from this process would have been the trigger for baleen-bearing whale evolution [43, 44, 53].

Paleoclimatic reconstructions based on $\delta^{18}O$ and Mg/Ca ratio depict the Late Oligocene as a warm period [79, 96, 131] (Figure 15). Zachos et al. [131]

stated that baleen-bearing whales appeared during the Early Oligocene (around 33 Ma) in concomitance with the strong glacial event associated to the ephemeral establishment of ice-sheet on Antarctica. The transient glacial regime in Antarctica was probably related to paleogeographic changes (i.e., the opening of the Tasman Sea and the Drake Passage) rather than to a true decline in temperature [96, 131]. The oceanographic and climatic situation of Early Oligocene was different from the warmer conditions of the latest part of this period when the Antarctic ice-sheet was probably completely decayed [131].

Starting from the basal Oligocene glaciation (Oi-1 event), wide portions of previously submersed continental shelves were exposed and submitted to strong erosion [110, 112]. The sedimentary flux provided the oceans with enormous amount of sialic nutrients together with Fe and Zn that enriched the waters stimulating high productivity [79, 116]. In particular, it has been documented that this process was not geographically restricted to the Antarctic waters but it involved also the whole continental margins of Africa and New Jersely [110, 112]. Other students reviewed the mechanisms of nutrient-intake and planktonic responses to changes in nutrient sources in modern waters providing an actualistic framework to understand the paleoecological evolution of ancient oceans [29, 75]. Based upon these works, it is possible to conclude that the Antarctic waters, the proto-Indo-Pacific, and the whole Atlantic Ocean became highly productive during the Late Oligocene being potentially able to sustain populations of baleen-bearing whales. The paleoenvironmental conditions of the proto-Mediterranean and the proto-Paratethys are not yet completely understood but the presence of eomysticetid and cetotheriid whales in Central Germany and Austria suggest that at least some portions of these basins were highly productive during the Late Oligocene.

In conclusion, it is likely that the tectonic-driven Oi-1 glaciation that triggered the Circum-Antarctic current was the key event that stimulated high productivity in the world oceans. However, due to the worldwide effects of this event, several places could be the center of origin of mysticetes because several oceans were productive enough to sustain population of baleen-bearing whales. Such an observation is further supported by the recent discovery of toothed mysticetes from the Late Oligocene of South Carolina [7].

Miocene: Dispersal

A massive pulse in diversity of baleen-bearing mysticetes is observed in the Miocene following the origin of the group and its early spreading across the Tethys (Figure 14). The high diversity of the Miocene is due to the origins of most of the modern-day families (Eschrichtiidae, Balaenidae, Balaenopteridae) and to the worldwide radiation of archaic mysticetes and the Cetotheriidae.

A number of families, genera and species have been reported from deposits in Italy, Germany, Belgium, Holland, Portugal, United States, Japan, Australia, New Zealand, South Africa, Peru. This wordwide distribution corresponded to high rates of endemicity. In general, genera are restricted to relatively small areas and very few contacts are documented between localities. In at least two cases, 'cetothere' genera are reported from the East Coast of the United States and Japan [73, 74] but these are the only cases of genus sharing by Miocene deposits separated by several thousand kilometers.

One factor influencing our perception of the endemicity of the mysticete fossil record from Miocene consists in the fact that most of the known taxa have been established in XIX and early in the XX century and are in critical need of revision. It may be that new analyses will reveal that some taxa have to be discarded and substituted by others based on distantly located types. However, after almost 14 years of study of fossil and living mysticetes I must confess that, in my opinion, it is highly unlikely that the current knowledge of the taxonomic compositions of mysticete faunas from distant basins will change significantly after a substantial revision of the fossil record. Fossil collections located in Belgium, Italy, United States, Holland, Japan really include different species that record lots of endemic taxa in the Miocene.

It is not easy to understand the process underlying the origin of endemic species in the pelagic realm. In general, it is assumed that mysticete species are worldwide distributed and that individual whales are able to move all around the world. On the contrary, living mysticetes live in discrete groups (stocks) whose feeding and reproductive grounds are substantially fixed and movements from one stock to another are rather rare [59]. Incomplete land barriers may also been sufficient to subdivide marine areals inhabited by mysticetes. The case of Gibraltar is quite exemplar: in the modern-day Mediterranean Sea, Gibraltar is the extreme western border of the Mediterranean basin but it does not separate totally the Mediterranean from the North Atlantic Ocean; it allows movements of mysticetes from outside the Mediterranean to inside it and *vice versa*. However, it is observed that the fin whales inhabiting the Mediterranean exit only occasionally and the Mediterranean population is becoming genetically isolated from the North Atlantic population [95]. This means that also in the pelagic realm conditions can exist that are able to limit mysticete movements promoting their permanence in relatively small basins. Taking this information in mind, it is quite easy to imagine that the fragmentation of the Tethys into a series of different basins in the Miocene could have led to the isolation of mysticete populations that became genetically isolated. This mechanism may work in a context where food abundance is enough to support permanent populations in small areas.

The radiation of Miocene Balaenomorpha followed the first glacial maximum of the Miocene (Mi-1 event of ref. [131]). It should be remembered that also the radiation of the earliest baleen-bearing mysticetes (occurred around 33-30 Ma) was contemporaneous to the warmer period that followed the glacial maximum of Early and middle Oligocene. The orbitally-driven Mi-1 event was particularly long (about 200 Ka) and was followed by a series of small glaciations. This event was also contemporaneous with the divergence of pinnipeds [9, 10, 11] and this suggests that it represented a major turning point in marine mammal evolution. Keller and Barron [61] stated that between 22 and 21 Ma a massive faunal turnover in benthic foraminifers was paralleled by the most significant erosive event of the whole Miocene (20-18 Ma). This event is related to an increase in productivity and predates the establishment of the present circulation pattern. The emergence of pinnipeds and advanced Cetotheriidae and the patterning of this new paleoecological situation are probably linked and it is highly likely that all of that was triggered by the Mi-1 glacial event of the Early Miocene.

The Late Miocene development of mysticete diversity was associated to a climate that was colder than that immediately following the Mi-1 event (Figure 15), a climate with declining temperature because of the definitive establishment of a large-scale ice-sheet on East Antarctica [131]. However, between 14 and 8 Ma, there was a strongly positive oscillation of $\delta^{13}C$ (maybe the highest since mid-Paleocene). High values of $\delta^{13}C$ were maintained up to 7.5 Ma. When the atmospheric pCO_2 is assessed, it appears that the concentration of this gas was approximately constant over the Miocene with two main exceptions: (1) an episode of rise, fall and subsequent rise of CO_2 occurred approximately between 16 and 15 Ma, and (2) a subsequent episode of carbon dioxyde increase at the end of Miocene around 6 Ma [96]. These episodes occurred in oceans whose pH values were steadily higher than those calculated for Paleogene [96].

The increase in $\delta^{13}C$ in the Late Miocene oceans should be linked to a massive input of nutrients. In fact, increase in carbon fractions means increase in nutrient availability [106]. Moreover, it has been demostrated that a correlation exists between $\delta^{13}C$ values and biodiversity since 455 Ma suggesting a strong connection between nutrient availability, diversity of land plants, and diversity in marine animals. As for the the early radiation of Oligocene and Early Miocene Balaenomorpha, the availability of more nutrients could have been the reason of the range expansion of mysticetes in the Late Miocene. However, while the early radiation occurred in warmer climate, the Late Miocene dispersal occurred during a period of declining temperature and after the non-ephemeral establishment of an ice cap on East Antarctica.

Pliocene-to-Recent: Origin of the Modern Mysticete Fauna

The mysticete diversity of the Pliocene is much lower than that observed in the Miocene and is surely influenced by the massive extinctions that led the archaic 'cetotheres to disappear and to a strong contraction of the diversity of Cetotheriidae *s.s.* In the Pliocene many new balaenid taxa appeared and quickly attained a worldwide distribution. *Balaena* and *Balaenula* are among them. Both are known from Italy, Belgium, United States (East Coast), and Japan. *Herpetocetus* is maybe the last surviving cetotheriid whale which is known to occur up to the middle Pliocene in Japan, Belgium, United States (East and West Coasts). The Mediterranean record shows the persistence of balaenopterids with archaic morphologies [25] up to the Middle Pliocene.

This high diversity is severly depleted by the end of the Middle Pliocene possibly by a series of coordinated extinction events occurred worldwide (Figure 15). It is still unclear the degree of contemporaneity of the extinctions of Belgian, Mediterranean, Californian, and Japanese *Balaenula* but it is quite clear that while these forms were living in the Early and early Middle Pliocene, none of them was still living in the Late Pliocene. Apart from the loss of biodiversity observed from the lower and the upper Pliocene, there is also another point that is worth considering. In the Early Pliocene, *Herpetocetus*, *Eschrichtioides*, balaenids, and balaenopterids were characterized by different trophic strategies and behaviours [20]. There was a certain amount of coexistence of balaenids, balaenopterids, latest cetotheriids and, at least somewhere, eschrichtiids in small basins such as the Japan Sea and the Mediterranean [20, 91]. In the Late Pliocene many of these trophic behaviours disappeared from most of the basins. It is still unclear if the diversity loss was due to disappearance of particular food items or if it was driven by other factors (*i.e.*, climate change) and it is not clear if a general restructuration of oceanic trophic webs led these mysticetes to the extinction or if it was the mysticete extinctions that triggered massive changes in the oceanic trophic webs. A precise calibration of mysticete extinctions in the Pliocene is necessary to understand the relationships between the diversity loss and the oceanic global changes.

Whitmore [124] suggested that the origin of the modern mysticete fauna is a result of the emergence of present-day oceanographic pattern in North Atlantic. Several workers linked the origin of the modern circulation system and climate of North Atlantic to the establishment of the Arctic ice-sheet and the closure of the Panamanian Seaway. Climate and oceanic circulation were highly related during Pliocene as suggested by analyses of microfossils and numerical experiments. In fact, in contrast to the Oi-1 event that was tectonically-driven and discoupled from temperature trends, some authors suggest that the Plio-Pleistocene glacial regime was linked to an effective decrease in temperature [131].

There is a body of evidence supporting an Early Pliocene warm climate in the northern hemisphere [13, 14, 15, 32, 33, 39, 54, 56, 57] (Figure 15). Such a period occurred when Balaenidae experienced their highest diversity peak. In the Mediterranean, marked extinction rates among molluscs, benthic foraminifers and fishes dated approximately 3 Ma showing what is considered as the first significant temperature decrease of the Pliocene [88, 89, 99, 100, 101, 111, 129]. A second extinction pulse is also found in Mediterranean mollusc assemblages around 2.5 Ma [1, 88, 101]; it was suggested that this extinction pulse was connected with the definitive establishment of an Arctic ice-sheet [58, 98]. These climatic changes were related to dramatic remodelling of paleocirculation patterns in North Atlantic than, in their turns, influenced the heat amount transported from Caribbean to northern areas by the Gulf Stream [40] and the geographic developments of the global routes followed by warmer waters in the whole northern hemisphere [87].

Due to these large-scale modifications of Pliocene climates and paleocirculation patterns, and due also to the presence of balaenopterids outside the Atlantic Ocean in the Pliocene, it should be considered the hypothesis that the link betwen paleoenvironmental conditions and the origin of the modern mysticete fauna were more global than hypothesized by Whitmore [124].

A connection between the strong changes in paleocirculation patterns and the closure of the Panamanian Seaway was proposed by several workers [8, 55, 87, 119, 123]. The biotic effects of this sequence of tectonic events were enormous and generated new distribution patterns in both marine and continental realms [76, 114, 126]. It seems that the closure of the Panamanian Seaway did not influence the immediate survival of the archaic mysticete forms.

In this attempt to analyse extinction dynamics of mysticetes, the Pleistocene record is of no help. In fact, there are only a few fossil and subforssil materials that can give some information. Among them there are good specimens referable to the gray whale *Eschrichtius robustus* from North Atlantic which testify that the distribution of this species encompassed also areals that today are not inhabited by this species. Pleistocene balaenopterid specimens are known from Italy [78] and other places (see The Paleobiology Database at http://www.paleodb.org/cgi-bin/bridge.pl) but are still to be formally described.

What we see in modern seas is the final result of a process of diversity loss started by the end of the Miocene. A process that led to the origin and distribution of a few gigantic genera and species in different marine basins of the world. However, there are still several things that we do not know about the origin of the modern mysticete populations. Our ignorance is at least about the following points: (1) the real abundance of these populations before the starting of the industrial whaling is not known (genetic studies suggest that fin and humpback whale stocks of the Pacific and bowhead whale stocks in the Antarctic were formed by many thousands of individuals supporting a hypothesis of extremely abundant prey items [20, 103], and (2) we are rather ignorant about the origin of the antitropical distribution of balaenids and balaenopterids even though genetic studies support the working hypothesis of Equator crossing during glacial periods by right whales [104]. Study and publication of Pleistocene records is critically needed to reconstruct the history of Balaenomorpha in the last few million years.

CONCLUSION

The evolutionary history of baleen-bearing whales is closely linked to global climatic and tectonic changes. Modifications in erosive regimes, temperatures, paleocirculation patterns and paleogeography have surely influenced the course of mysticete evolution by making nutrients available and by providing conditions to support mysticete populations in small or large basins.

Even if the main climatic modifications of the Cenozoic appear deciphered, we are still ignorant about the precise influence of each event in the selective regimes affecting mysticete populations and it is not clear wether the morphological and taxonomic changes observed in mysticete faunas reflect adaptive trends generated by global or local environmental changes.

In the end, only a few gigantic and highly specialized species have survived up to the modern-day oceans and their further survival is under question because of human activities. Efforts directed at discovering phylogenetic relationships, paleobiogeographic patterns, and paleobiological characters of their ancestors will be more and more important in shaping shaper conservation strategies for these giants of the sea and will help us to be fully aware of the natural history of these slowly evolving, highly sophisticated marine mammals.

ACKNOWLEDGMENTS

This paper derives in part from my PhD dissertation at the University of Pisa. I am grateful to Frank Columbus for inviting me to participate in the preparation of this volume. I want to thank Albert E. Sanders (The Charleston Museum) and Lawrence G. Barnes (both my PhD commission) very much because of their thorough reviews of an early version of this paper as it was incorporated in my PhD thesis.

REFERENCES

[1] Allmon, W. D. (1992). Role of temperature and nutrients in extinction of turritelline gastropods: Cenozoic of the northwestern Atlantic and northeastern Pacific. *Palaeogeography Palaeoecology Palaeoclimatology*, 92, 41-54.

[2] Andrews, R. C. (1914). Monographs of the Pacific Cetacea – 1. The California Gray Whale (*Rhachianectes glaucus* Cope). *Memoirs of the American Museum of Natural History*, 1(5), 1-287.

[3] Arnason, U. and Gullberg A., (1994). Relationship of baleen whales established by cytochrome *b* gene sequence comparison. *Nature*, 367, 726-728.

[4] Bajpai, S. and Gingerich, P. D. (1998). A new Eocene archaeocete (Mammalia, Cetacea) from India and the time of origin of whales. *Proceedings of the National Academy of Science USA*, 95, 15464-15468.

[5] Baker, A. N. (1985). Pygmy right whale *Caperea marginata* (Gray, 1846). In S. H. Ridgway and R. Harrison, R. (Eds.), Handbook of marine mammals: The sirenians and baleen whales (Volume 3, pp. 345-354). New York, Academic Press.

[6] Barnes, L. G. and McLeod, S. A. (1984). The fossil record and phyletic relationships of gray whales. In M. L. Jones, S. Leatherwood, S. Swartz (Eds), *The gray whale* (pp. 3-32). Orlando, Academic Press.

[7] Barnes, L. G. and Sanders, A. E. (1996). The transition from Archaeocetes to Mysticetes: Late Oligocene toothed mysticetes from near Charleston, South Carolina. In J. E. Repetski (Ed.), Abstracts of papers: Sixth North American Paleontological Convention, (Paleontological Society, Special Publication, 8, p. 24). Washington, Smithsonian Institution.

[8] Barron, E. J. and Peterson, W. H. (1991). The Cenozoic ocean circulation based on ocean General Circulation Model results. *Palaeogeography Palaeoecology Palaeoclimatology*, 83, 1-28.

[9] Berta, A. (1994). A new species of phocoid pinniped *Pinnarctidion* from the Early Miocene of Oregon. *Journal of Vertebrate Paleontology*, 14, 405-413.

[10] Berta, A. and Adam, P. J. (2001). Evolutionary biology of pinnipeds. In J.-M. Mazin and V. De Buffrenil (Eds.), Secondary adaptations of tetrapods to life in water (pp. 235-260). München, Verlag Pfeil.

[11] Berta, A., Ray, C. E., Wyss, A. R. (1989). Skeleton of the oldest known pinniped *Enaliarctos mealsi*. *Science,* 244, 60-63.

[12] Berta, A. and Sumich, J. L. (1999). *Marine mammals: Evolutionary biology*. New York, Academic Press.

[13] Billups, K., Ravelo, A. C., Zachos, J. C. (1998). Early Pliocene deep water circulation in the western equatorial Atlantic: implications for high-latitude climate change. *Paleoceanography*, 13, 84-95.

[14] Billups, K., Ravelo, A. C., Zachos, J. C. (1998). Early Pliocene climate: a perspective from the western equatorial Atlantic warm pool. *Paleoceanography*, 13, 459-470.

[15] Billups, K., Ravelo, A. C., Zachos, J. C., Norris, R. D. (1999). Link between oceanic heat transport, thermohaline circulation, and the Intertropical Convergence Zone in the early Pliocene Atlantic. *Geology,* 27, 319-322.

[16] Bisconti, M. (2000). New description, character analysis and preliminary phyletic assessment of two Balaenidae skulls fromthe Italian Pliocene. *Palaeontographia Italica*, 87, 37-66.

[17] Bisconti, M. (2001). Morphology and postnatal growth trajectory of rorqual petrosal. *Italian Journal of Zoology*, 68, 87-93

[18] Bisconti, M. (2003). Evolutionary history of Balaenidae. *Cranium*, 20, 9-50.

[19] Bisconti, M. (2005). Skull morphology and phylogenetic relationships of a new diminutive balaenid from the Lower Pliocene of Belgium. *Palaeontology*, 48, 793-816.

[20] Bisconti, M. (2005). Paleontologia e conservazione: il caso della balena della Groenlandia. In F. Scapini (Ed.), La logica dell'evoluzione dei viventi: Spunti di riflessione. (Atti del XII Convegno del Gruppo Italiano di Biologia Evoluzionistica, pp. 133-142). Firenze, Firenze University Press.

[21] Bisconti, M. (2006). *Titanocetus*, a new baleen whale from the Middle Miocene of northern Italy (Mammalia, Cetacea, Mysticeti). *Journal of Vertebrate Paleontology*, 26, 344-364.

[22] Bisconti, M. (2007). A new basal balaenopterid whale from the Pliocene of northern Italy. *Palaeontology*, 50, 1103-1122.

[23] Bisconti, M. (2008). Morphology and phylogenetic relationships of a new eschrichtiid genus (Cetacea: Mysticeti) from the Early Pliocene of northern Italy. *Zoological Journal of the Linnean Society*, 153, 161-186.

[24] Bisconti, M. (2008). Le culture degli altri animali. Bologna, Zanichelli.

[25] Bisconti, M. (2009). Mediterranean record of balaenopterid evolution. In D. Schwartz-Wings, O. Wings, F. Sattler (Eds.), 7th Annual Meeting of the European Association of Vertebrate Paleontologists, Berlin. (Abstract Volume, p. 13). Aachen, Shaker Verlag.

[26] Bisconti, M. and Varola, A. (2006). Functional hypothesis on an unusual mysticete dentary with double coronold process from the Miocene of Apulia and its systematic and behavioural implications. *Palaeoncographia Italica* 87, 19-35.

[27] Bouetel, V. and DeMuizon, C. (2006). The anatomy and relationships of *Piscobalaena nana* (Cetacea, Mysticeti), a Cetotheriidae s.s. from the early Pliocene of Peru. *Geodiversitas*, 28, 319-395.

[28] Brandt, J. F. (1873). Untersuchungen uber die fossilen und subfossilen Cetaceen Europas. Memoires de l'Academie imperiale des Sciences, St. Pétersburg, 7, 1-54.

[29] Butler, A. (1998). Acquisition and utilization of transition metal ions by marine organisms. *Science*, 281, 189-243.

[30] Cabrera, A. (1926). Cetaceos fosiles del Museo de La Plata. *Revista del Museo de la Plata*, 29, 363-411.

[31] Cande, S. C., Stock, J. M., Muller, D., Ishihara, T. (2000). Cenozoic motion between East and West Antarctica. *Nature*, 404, 145-150.

[32] Chandler, M., Rind, D., Thompson, R. (1994). Joint investigations of the middle Pliocene climate II: GISS GCM Northern Hemisphere results. *Global and Planetary Change*, 9, 197-219.

[33] Cronin, T. M., Whatley, R., Wood, A., Tsukagoshi, A., Ikey, A. N., Brouwers, E.M., Briggs, Jr W. M. (1993). Microfaunal evidences for elevated Pliocene temperatures in the Arctic Ocean. *Paleoceanography*, 8, 161-173.

[34] Dathe, F. (1983). *Megaptera hubachi* n. sp., ein fossiler Bartenwal aus marinen Sandsteinschichten des tieferen Pliozäns Chiles. Zeitschrift für Geologische Wissenschaften, 11, 813-852.

[35] Deméré, T. A. and Berta, A. (2008). Skull anatomy of the Oligocene toothed mysticete *Aetioceus weltoni* (Mammalia; Cetacea): implications for mysticete evolution and functional anatomy. *Zoological Journal of the Linnean Society*, 154, 308-352.

[36] Deméré, T. A., Berta, A., McGowen, M. R. (2005). The taxonomic and evolutionary history of fossil and modern balaenopteroid mysticetes. *Journal of Mammalian Evolution*, 12, 99-143.

[37] Deméré, T. A. and Cerutti, R. (1982). A Pliocene shark attack on a cetotheriid whale. *Journal of Paleontology*, 56, 1480-1482.

[38] Deméré, T. A., McGowen, M. R., Berta, A., Gatesy, J. (2008). Morphological and molecular evidence for a stepwise evolutionary transition from teeth to baleen in mysticete whales. *Systematic Biology*, 57, 15-37.

[39] Dowsett, H., Thompson, R., Barron, J., Cronin, T., Fleming, F., Ishman, S., Poore, R., Willard, D., Holtz, Jr T. (1994). Joint investigations of the middle Pliocene climate I: PRISM paleoenvironmental reconstructions. *Global and Planetary Change*, 9, 169-195.

[40] Edwards, L. E., Mudie, P. J., De Vernal, A. (1991). Pliocene paleoclimatic reconstruction using dinoflagellate cysts: comparison of methods. *Quaternary Science Reviews*, 10, 259-274.

[41] Fitzgerald, E. M. G. (2004). A review of the Tertiary fossil Cetacea (Mammalia) localities in Australia. *Memoirs of Museum Victoria*, 61, 183-208.
[42] Fitzgerald, E. M. G. (2006). A bizarre new toothed mysticete (from Australia) and the early evolution of baleen whales. Proceedings of the Royal Society B: *Biological Sciences*, 273, 2955-2963.
[43] Fordyce, R. E. (1977). The development of the circum-Antarctic current and the evolution of the Mysticeti (Mammlia: Cetacea). *Palaeogeography Palaeoecology Palaeoclimatology*, 21, 265-271.
[44] Fordyce, R. E. (1980). Whale evolution and Oligocene southern ocean environments. *Palaeogeography Palaeoecology Palaeoclimatology*, 31, 319-336.
[45] Fordyce, R. E. (1984). Evolution and zoogeography of cetaceans in Australia. In M. Archer and G. Clayton (Eds.), Vertebrate zoogeography and evolution in Australasia (pp. 929-948). Perth, Hesperion.
[46] Fordyce, R. E. (2002). Oligocene origins of skim-feeding right whales: a small, archaic balaenid from New Zealand. *Journal of Vertebrate Paleontology,* 22(3Suppl.), 54.
[47] Fordyce, R. E. (2006). An unexpected diversity of basal baleen whales: Late Oligocene Eomysticetidae from New Zealand. *Journal of Vertebrate Paleontology*, 26, Suppl., 62A.
[48] Fordyce, R. E. and Barnes, L. G. (1994). The evolutionary history of whales and dolphins. *Annual Review of Earth and Planetary Science*, 22, 419-455.
[49] Fordyce, R. E. and DeMuizon C. (1999). Evolutionary history of cetaceans: a review. In J.-M. Mazin and V. De Buffrenil (Eds.), Secondary adaptations of tetrapods to life in water (pp. 169-233). München, Verlag Pfeil.
[50] Gaskin, D. E. (1986). The ecology of whales and dolphins. London, Heinemann Educational Books.
[51] Geisler, J. H. and Luo, Z. (1996). The petrosal and inner ear of Herpetocetus sp. (Mammalia: Cetacea) and their implications for the phylogeny and hearing of archaic mysticetes. *Journal of Paleontology*, 70, 1045-1066.
[52] Geisler J. H. and Sanders, A. E. (2003). Morphological evidence for the phylogeny of Cetacea. *Journal of Mammalian Evolution*, 10, 23-129.
[53] Gingerich, P. D. (2005). Cetacea. In K. D. Rose and J. D. Archibald (Eds.), Placental mammals: origin, timing, and relationships of the major extant clades (pp. 234-252). Baltimore, Johns Hopkins University Press.

[54] Groot, J. J. (1991). Palynological evidence for Late Miocene, Pliocene and Early Pleistocene climate changes in the middle U.S. Atlantic Coastal Plain. *Quaternary Science Reviews*, 10, 147-162.
[55] Haug, G. H. and Tiedemann R. (1998). Effect of the formation of the Isthmus of Panama on Atlantic Ocean thermohaline circulation. *Nature*, 393, 673-676.
[56] Haywood, A. M., Sellwood, B. W., Valdes, P. J. (2000). Regional warming: Pliocene (3 Ma) paleoclimate of Europe and the Mediterranean. *Geology*, 28, 1063-1066.
[57] Haywood, A. M., Valdes, P. J., Sellwood, B. W. (2000). Global scale palaeoclimate reconstruction of the middle Pliocene climate using the UKMO GCM: initial results. *Global and Planetary Change*, 25, 239-256.
[58] Hodell, D. A. and Warnke, D. A. (1991). Climatic evolution of the Southern Ocean during the Pliocene epoch from 4.8 to 2.6 million years ago. *Quaternary Science Reviews*, 10, 205-214.
[59] Hoelzel, A. R. (1994). Genetics and ecology of whales and dolphins. *Annual Review of Ecology and Systematics*, 25, 377-399.
[60] Huelsenbeck, J. P. (1994). Comparing the stratigraphic record to estimates of phylogeny. *Paleobiology,* 20, 470-483.
[61] Keller, G. and Barron, J. A. (1983). Paleoceanographic implications of Miocene deep-sea hiatuses. *Geological Society of America Bulletin*, 94, 590-613.
[62] Kellogg, R. (1922). Description of the skull of *Megaptera miocaena*, a fossil humpback whale from the Miocene diatomaceous earth of Lompoc; California. Proceedings of the United States National Museum, 61, 1-18.
[63] Kellogg, R., (1924). Description of a n4ew genus and species of whalebone whales from the Calvert Cliffs, Maryland. *Proceedings of the United States National Museum*, 63, 1-14.
[64] Kellogg, R. (1934). The Patagonian fossil whalebone whale Cetotherium moreni (Lydekker). Contributions to Palaeontology, Carnegie Institution, Washington, 447, 65-81.
[65] Kellogg, R. (1934). A new cetothere from the Modelo Formation at Los Angeles, California. Contributions to Palaeontology, Carnegie Institution, Washington, 447, 85-104.
[66] Kellogg, R. (1965). A new whalebone whale from the Miocene Calvert Formation. United States National Museum, Bulletin, 247, 1-45.

[67] Kellogg, R. (1968). Miocene Calvert Mysticetes described by Cope. United States National Museum, Bulletin, 247, 103-132.

[68] Kellogg, R. (1968). A hitherto unrecognized Calvert cetothere. United States National Museum, Bulletin, 247, 133-161.

[69] Kellogg, R. (1968). A sharp-nosed cetothere from the Miocene Calvert. United States National Museum, Bulletin, 247, 163-173.

[70] Kellogg, R. (1968). Supplement to description of Parietobalaena palmeri. United States National Museum, Bulletin, 247, 175-197.

[71] Kimura, T. (2006). Feeding strategy of an Early Miocene cetothere from the Toyama and Akeyo Formations, central Japan. *Paleontological Research*, 6, 179-189.

[72] Kimura, T. and Ozawa, T. (2002). A new cetothere (Cetacea: Mysticeti) from the Early Miocene of Japan. *Journal of Vertebrate Paleontology*, 22, 684-702.

[73] Kimura, T., Sakamoto, O., Hasegawa, Y. (2000). A Middle Miocene cetothere from the Chichibumachi Group, Central Japan. *Bulletin of the Saitama Museum of Natural History,* 18, 15-29.

[74] Kimura, T., Sakamoto, O., Hasegawa, Y. (1998). A cetothere from the Miocene Chichibumachi Group, Aitama Prefecture. Japan. *Bulletin of the Saitama Museum of Natural History,* 16, 1-13.

[75] Jickells, T. D. (1998). Nutrient biogeochemistry of the coastal zone. *Science*, 281, 217-221.

[76] Ibaraki, M. (1997). Closing of the Central American Seaway and Neogene coastal upwelling along the Pacific coast of South America. Tectonophysics, 281, 99-104.

[77] Ichishima, H, Sato, E, Sagayama, T, Kimura, M. (2006). The oldest record of Eschrichtiidae (Cetacea: Mysticeti) from the Late Pliocene, Hokkaido, Japan. *Journal of Paleontology*, 80, 367-379.

[78] Landini W., Bianucci G., Bisconti M., Carnevale G., Sorbini C. (2005). I vertebrati marini. In L. Bonfiglio (Ed.), Paleontologia dei Vertebrati in Italia: Evoluzione biologica, significato paleoambientale e paleogeografia. Memorie del Museo Civico di Storia Naturale di Verona, 2. Serie, Sezione Scienze della Terra, vol. 6, 171-182.

[79] Lear, C. H., Elderfield, H., Wilson, P. A. (2000). Cenozoic deep-sea temperatures and global ice volumes from Mg/Ca in benthic foraminiferal calcite. *Science*, 287, 269-272.

[80] May-Collado, L. and Agnarsson, I. (2006). Cytochrome *b* and Bayesian inference of whale phylogeny. *Molecular Phylogenetics and Evolution*, 38, 344-354.

[81] Mcleod, S. A., Whitmore, Jr F. C., Barnes, L. G. (1993). Evolutionary relationships and classification. In J. J. Burns, J. J. Montague, C. J. Cowles (Eds.), The bowhead whale. The Society for Marine Mammalogy, Special Publication, 2, 45-70.

[82] Mead J. G. and Mitchell, E. D. (1984). Atlantic Gray Whales. In M. L. Jones, S. Leatherwood, S. Swartz (Eds.), The Gray Whale (pp. 33-53). New York, Academic Press.

[83] Messenger, S. and Mcguire, J. A. (1998). Morphology, molecules, and the phylogenetics of cetaceans. *Systematic Biology*, 47, 90-124.

[84] Milinkovitch, M. C., Ortí, G., Meyer, A. (1993). Revised phylogeny of whales suggested by mitochondrial ribosomal DNA sequences. *Nature*, 361, 346-348.

[85] Milinkovitch, M., Meyer, A., Powell, J. R. (1994). Phylogeny of all major groups of cetaceans based on DNA sequences from three mitochondrial genes. *Molecular Biology and Evolution*, 11, 939-948.

[86] Mitchell, E. D. (1989). A new cetacean from the Late Eocene La Meseta Formation, Seymour Island, Antarctic Peninsula. *Canadian Journal of Fisheries and Aquatic Science*, 46, 2219-2235.

[87] Molina-Cruz, A. (1997). Closing of the Central American Gateway and its effects on the distribution of Late Pliocene radiolarians in the eastern tropical Pacific. *Tectonophysics*, 281, 105-111.

[88] Monegatti, P and Raffi, S. (2001). Taxonomicc diversity and stratigraphic distribution of Mediterranean Pliocene bivalves. *Palaeogeography, Palaeoecology, Palaeoclimatology*, 165, 171-193.

[89] Morely, J. J. and Dworetzky, B. A. (1991). Evolving Pliocene-Pleistocene climate: a North Pacific perspective. *Quaternary Science Reviews*, 10, 225-238.

[90] Noad, M. J., Cato, D. H., Bryden, M. M., Jenner, M.-N., Jenner, K. C. S. (2000). Cultural revolution in whales songs. *Nature*, 408, 537.

[91] Oishi, M and Hasegawa, Y. (1994). A list of fossil cetaceans from Japan. *The Island Arc*, 3, 493-505.

[92] Oishi, M., Ono, K., Kawakami, T., Sato, J., Nokariya, H., Hasegawa, Y. (1985). Pliocene baleen whales and bony-toothed bird from Iwate Prefecture, Japan (Parts I-VI). *Bulletin of the Iwate Prefectural Museum*, 3, 143-162.

[93] Okazaki, Y. (1994). A new type of primitive baleen whale (Cetacea: Mysticeti) from Kyushu, Japan. *The Island Arc*, 3, 432-435.

[94] Packard, E. L. and Kellogg, R. (1934). A new cetothere from the Miocene Astoria Formation of Newport, Oregon. Contributions to Palaeontology, Carnegie Institution Washington, 447, 3-81.

[95] Palsbøll, P. J., Bérubé, M., Aguilar, A., Notarbartolo di Sciara, G., Nielsen, R. (2004). Discerning between recurrent gene flow and recent divergence under a finite-site mutation model applied to north atlantic and mediterranean sea fin whale (*Balaenoptera physalus*) populations. *Evolution*, 58, 670-675.

[96] Pearson, P. N. and Palmer, M. R. (2000). Atmospheric carbon dioxyde concentrations over the past 60 million years. *Nature*, 406, 695-699.

[97] Pilleri, G. (1986). Beobachtungen an den fossilen Cetaceen des Kaukasus. Vammala, Hirnanatomisches Institut Ostermundigen.

[98] Prentice, M. L. and Denton, G. H. (1988). The deep-sea Oxygen isotope record, the global ice sheet system and hominid evolution. In F.E. Grine (Ed.), Evolutionary history of the "robust" australopithecines (pp. 383-403). New York, De Gruyter.

[99] Raffi, S. (1993). L'origine della fauna attuale del Mar Mediterraneo: il caso dei bivalvi. Contributi del Centro Linceo Interdisciplinare "Beniamino Segre", 86, 189-207.

[100] Raffi, S. and Monegatti, P. (1993). Bivalve taxonomic diversity throughtout the Italian Pliocene as a tool for climatic-oceanographic and stratigraphic inferences. *Ciencias da Terra,* 12, 45-50.

[101] Raffi, S., Stanley, S. M., Marasti, R. (1985). Biogeographic patterns and Plio-Pleistocene extinction of Bivalvia in the Mediterranean and southern North Sea. *Paleobiology*, 11, 368-389.

[102] Ridgway, S. M. and Harrison, R. (1985). Handbook of marine mammals. London, Academic Press.

[103] Roman, J. and Palumbi, S. R. (2003). Whales before whaling in the North Atlantic. *Science,* 301, 508-510.

[104] Rosembaum, H. C., Brownell, Jr R. L., Brown, M. W., Schaeff, C., Portway, V., White, B. N., Malik, S., Pastene, L. A., Patenaude, N. J., Baker, C. S., Rowntree, V., Tynan, C. T., Bannister, J. L., DeSalle, R. (2000). World-wide genetic differentaition of *Eubalaena*: questioning the number of right whale species. *Molecular Ecology*, 9, 1793-1802.

[105] Rothausen, K. (1971). *Cetotheriopsis tobieni* n. sp., der erste Paläogene Bartenwal (Cetotheriidae, Mysticeti, Mamm.) nördlich des Tethysraumes. Abhandlungen des Hessischen Landesamtes für Bodenforschung, 60, 131-148.

[106] Rothman, D. H. (2001). Global biodiversity and the ancient carboncycle. *Proceedings of the National Academy of Science USA*, 98, 4305-4310.

[107] Sanders, A.E. and Barnes, L.G. (2002). Paleontology of the Late Oligocene Ashley and Chandler Bridge Formations of South Carolina, 2: *Micromysticetus rothauseni*, a primitive cetotheriid mysticete (Mammalia: Cetacea). *Smithsonian Contributions to Paleobiology*, 93, 271-293.

[108] Sanders, A. E. and Barnes, L. G. (2002). Paleontology of the Late Oligocene Ashley and Chandler Bridge formations of South Carolina 3: Eomysticetidae, a new family of primitive Oligocene mysticetes (Mammalia: Cetacea), from South Carolina, U.S.A. *Smithsonian Contributions to Paleobiology*, 93, 313-356.

[109] Sanderson, L. R. and Wassersug, R. (1993). Convergent and alternative designs for vertebrate suspension feeding. In J. Hanken and B. K. Hall (Eds.), The skull: Functional and evolutionary mechanisms (Volume 3, pp. 37-112). Chicago, University Press of Chicago.

[110] Seranne, M. (1999). Early Oligocene stratigraphic turnover on the west Africa continental margin: a signature of the Tertiary greenhouse-to-icehouse transition? *Terra Nova*, 11, 135-140.

[111] Sprovieri, T. (1985). Paleotemperature changes and speciation among benthic Foraminifera in the Mediterranean Pliocene. Bollettino della Società Paleontologica Italiana, 24, 13-21.

[112] Steckler, M. S., Reynolds, D. J., Coakley, B. J., Swift, B. A., Jarrard, R. (1993). Modelling passive margin sequence stratigraphy. In H. W. Posamentier (Ed.), Sequence Stratigraphy and Facies Associations. International Association of Sedimentology, Special Publication, 18, 19-41.

[113] Steeman, M. E. (2007). Cladistic analysis and a revised classification of fossil and recent mysticetes. *Zoological Journal of the Linnean Society*, 150, 875-894.

[114] Stehli, F. G. and Webb, S. D. (1985). The great American biotic interchange. New York, Plenum Press.

[115] Stewart, B. S. and Leatherwood, S. (1985). Minke whale – *Balaenoptera acutorostrata*.. In S. M. Ridgway and R. Harrison (Eds.) Handbook of marine mammals: the sirenians and baleen whales (Volume 3, pp. 91-136). London, Academic Press.

[116] Szalamy, K. A. and Zachos, J. C. (1999). Late Eocene-Early Oligocene Climate Change on Southern Ocean Fertility: Inferences from Sediment

Accumulation and Stable Isotope Data. *Palaeogeogeography Palaeoclimatology Palaeoecology*, 145, 79-93.

[117] Tomilin, A. G. (1967). Cetacea. Israel Program for Scientific Translations, Jerusalem, 9, 1-756 pp.

[118] True, F. W. (1904). The whalebone whales of the western north Atlantic. *Smithsonian Contributions to Knowledge*, 33, 1-322.

[119] Tsuchi, R. (1997). Marine climatic responses to Neogene tectonics of the Pacific Ocean seaways. *Tectonophysics,* 281, 113-124.

[120] Uhen, M. D., Fordyce, R. E., Barnes, L. G. (2007). Mysticeti. In C. M. Janis, G. F. Gunnell, M. D. Uhen (Eds.), Evolution of Tertiary Mammals of North America (pp. 607-628). Cambridge, Cambridge University Press.

[121] Van Beneden, P.-J. (1872). Les baleines fossiles d'Anvers. *Bulletin de l'Academie Royale du Belgique,* 34, 6-20.

[122] Wada, S., Oishi, M., Yamada, T. K. (2003). A newly discovered species of living baleen whale. *Nature,* 426, 278-281.

[123] Wayne-Nesbitt, H. and Young, G. M. (1997). Sedimentation in the Venezuelan Basin, circulation in the Caribbean Sea, and onset of northern hemisphere glaciation. *The Journal of Geology*, 105, 531-544.

[124] Whitmore, Jr F. C. (1994). Neogene climatic change and the emergence of the modern whale fauna of the North Atlantic Ocean. *Proceedings of the San Diego Society of Natural History*, 29, 223-227.

[125] Whitmore, Jr F. C. and Barnes, L. G. (2008). The Herpetocetinae, a new subfamily of extinct baleen whales (Mammalia, Cetacea, Cetotheriidae). *Virginia Museum of Natural History Special Publication*, 14, 141-180.

[126] Whitmore, F. C. and Stewart, R. H. (1965). Miocene mammals and Central American Seaways. *Science*, 148, 180-185.

[127] Wiens, J. J. (2003a). Missing data, incomplete taxa, and phylogenetic accuracy. *Systematic Biology*, 52, 528-538.

[128] Wiens, J. J. (2003b). Incomplete taxa, incomplete characters, and phylogenetic accuracy: is there a missing data problem? *Journal of Vertebrate Paleontology*, 23, 297-310.

[129] Winkler, A., Wolf-Welling, T. C. W., Statteger, K., Thiede, J. (2002). Clay mineral sedimentation of high northern latitude deep-sea basins since the Middle Miocene (ODP Leg 151, NAAG). *International Journal of Earth Sciences*, 91, 133-148.

[130] Wolman, A. A. (1985). Gray whale – *Eschrichtius robustus*. In S. M. Ridgway and R. Harrison (Eds.), Handbook of marine mammals: The

sirenians and baleen whales (Volume 3, pp. 67-90). London, Academic Press.

[131] Zachos, J., Pagani, M., Sloan, L., Thomas, E., Billups, K. (2001). Trends, rhythms, and aberrations in global climate 65 Ma to present. *Science*, 292, 686-693.

[132] Zeigler, C. V., Chan, G. L., Barnes, L. G. (1997). A new Late Miocene balaenopterid whale (Cetacea: Mysticeti), Parabalaenoptera baulinensis, (new genus and species) from the Santa Cruz Mudstone, Point Reyes Peninsula, California. *Proceedings of the California Academy of Sciences,* 50, 115-138.

In: Whales and Dolphins
Editors: Craig A. Murray, pp. 47-88
ISBN: 978-1-61668-693-2

Chapter 2

WHALES AND CLIMATE CHANGE

Mark J. Spalding
The Ocean Foundation, Washington, DC

Near opposite poles on earth in their respective summers, an amazing process and journey occurs that has happened for thousands of year. Whales travel through icy waters, seeking copious amounts of food to nourish their massive bodies which are uniquely suited for their extreme environments.

Following the Arctic ice edge, navigating under thick ice or surfacing in open water or polynyas[1], the bowhead whale makes its annual journey similar to its ancestors countless generations before. It traces the path to the eastern Beaufort Sea where the rich waters team with food to nourish this thick skulled baleen whale. Constantly eating as it slowly swims, the bowhead skims the water for its desired meal of plankton and tiny crustraceans like krill, copepods and pteropods using its long and fine baleen filters.

In the Southern Ocean during the Antarctic summer, the blue whale lunges into dense groups of tiny krill, copepods and plankton. The planet's largest animal gulps huge mouthfuls by the ton as its throat pleats expand to form a gigantic food-laden pouch and then it forces water through hanging baleen plates which act as a sieve to catch the bounty.

---

[1] Polynyas are open water locations within the ice pack that are almost always clear of ice, caused by persistent physical conditions such as upwellings or wind patterns. They are important areas where marine mammals can surface to breathe, and can also provide productive feeding areas for bowheads, belugas, and narwhals. (Laidre & Heide-Jørgensen 2005).

The process of these large animals eating some of the smallest, occurs when their ecosystems are in balance. In simplest terms, for abundant food availability, the system needs the appropriate mix of sea ice, water temperature, currents, salinity, and sun light. These systems and the whales themselves are in serious danger and the culprit originates far from their pristine waters. Global climate change has begun to alter their habitats and those of whales in more temperate regions with documented adverse impacts already seen. What exact changes will occur is not certain but the outlook is especially grim unless substantial reductions in greenhouse gas emissions happen in short order.

The Changing Climate

The occurrence of global warming as a result of human-created greenhouse gas emissions - carbon dioxide, methane, and nitrous oxides – is no longer subject to credible scientific dispute. In a bleak and powerful assessment of the planet's future, the leading international network of climate scientists concluded that global warming is "unequivocal" and that human activity is the main driver, "very likely" causing most of the rise in temperatures since 1950 (IPCC 2007a). In February 2007, the Intergovernmental Panel on Climate Change (IPCC)[2] released its "Fourth Assessment Report" on the causes and consequences of climate change. For the first time, the IPCC asserted with near certainty – more than 90% confidence – that carbon dioxide and other greenhouse gases from human activities have caused warming in the past fifty years.[3] While there is continuing debate about the rate or regional effects of global warming and policy debates about what should be done to address the issue, the best available science clearly demonstrates that global warming exists and will bring profound changes to the world's climate. On the hopeful side, however,

[2] In 1988, the World Meteorological Organization and the United Nations Environment Programme established the IPCC. Its mission is to assess available scientific and socio-economic information on climate change and its impacts and the options for mitigating climate change and to provide, on request, scientific and technical advice to the Conference of the Parties to the United Nations Framework Convention on Climate Change. Since 1990, the IPCC has produced a series of reports, papers, methodologies, and other products that have become the standard works of reference on climate change.

[3] The Report is the product of some 600 authors from 40 countries. Over 620 expert reviewers and a large number of government reviewers also participated. Representatives from 113 governments, including the United States, reviewed and revised the summary line-by-line before adopting it and accepting the underlying report.

the IPCC report said warming and its harmful consequences could be substantially blunted by prompt action.

Over 400,000 years before modern industrialization, atmospheric carbon dioxide (CO_2) concentrations fluctuated between a relatively narrow range of 200 – 280 parts per million (ppm) (Doney, 2006). However, greenhouse gas emissions after industrialization have caused atmospheric CO_2 concentrations to move toward 380 ppm (Feely *et al.*, 2004; Raven *et a.l*, 2005); there is now approximately thirty percent more atmospheric CO_2 than in the 1700s (Doney, 2006). By 2030, CO_2 emissions may be fifty percent higher if current trends continue (Turley *et al.*, 2006) and by the century's end, atmospheric CO_2 concentrations may reach over 800 ppm according to Intergovernmental Panel on Climate Change emission scenarios (Prentice *et al.*, 2001).

Climate change has the potential to profoundly change high latitude ecosystems (Ragen et al., 2008). Temperature and sea level rise will lead to coastal and interior flooding (McCarthy et al. 2001). Sea ice habitat (Parkinson et al. 1999), will disappear in many areas and change where it continues to exist. Sea ice is projected to shrink in both the Arctic and Antarctic under all future emission scenarios (IPCC 2007a). This will cause extensive redistribution of mobile species, the disappearance of non-mobile species throughout portions of their range, and possible species extinction (Thomas et al. 2004). Climate change will modify water and nutrient cycles and energy pathways in the world's oceans (Macdonald et al. 2005). It may change oceanic and atmospheric circulation patterns (Schmittner 2005). All of these impacts are likely to cause severe biological and ecological consequences for marine and terrestrial ecosystems, with cascading changes through physical, chemical, and biological linkages (Ragen 2008). Many unexpected events have occurred already such as the rapid disintegration of the Larsen B ice shelf in the Antarctic (Shepherd et al. 2003), the ice-free Arctic Northwest passage and the swift glacial decline on Greenland (Rignot and Kanagaratnam 2006). These earlier-than-anticipated changes likely foreshadow many future surprises regarding climate change's regional and global effects (Ragen et al. 2008).

Climate change in the high latitude marine environment has several physical manifestations. These include changes in air, sea surface and water column temperatures, sea ice extent and thickness, terrestrial ice, precipitation, freshwater flow and salinity, sea level, and oceanic and atmospheric circulation (Ragen et al.2008, ACIA 2004, Walsh 2008).

Arctic Climate Change

The polar regions are disproportionately affected by higher levels of warming. In November 2004, the Arctic Climate Impact Assessment (ACIA)[4] released its report on "Impacts of a Warming Arctic." The report concluded that greenhouse gas driven climate changes "are being experienced particularly intensely in the Arctic (ACIA 2004). Arctic average temperature has risen at almost twice the rate as the rest of the world in the past 100- years (IPCC, 2007a). Widespread melting of glaciers and sea-ice and rising permafrost temperatures present additional evidence of strong arctic warming" (ACIA 2004). Significantly, ACIA reports that "acceleration of these climatic trends is projected to occur during this century, due to ongoing increases in concentrations of greenhouse gases in the earth's atmosphere."

In Alaska and western Canada, winter temperatures have increased by as much as 3-4 degrees Celsius (C), 5-7 degrees Fahrenheit (F) in the past 50 years. Over the next 100 years, under a moderate emissions scenario, annual average temperatures are projected to rise 3-5 degrees C (5-9 degrees F) over land and up to 7 degrees C (13 degrees F) over the oceans. Winter temperatures are projected to rise by 4-7 degrees C (5-9 degrees F) over land and 7-10 degrees C (13-18 degrees F) over the oceans (ACIA 2004).

This ongoing and projected warming has already and will continue to severely reduce the extent of sea-ice coverage. Over the past 30 years, the average sea-ice extent has decreased by about 8% each year, or nearly one million square kilometers (386,000 square miles), an area larger than all of Norway, Sweden, and Denmark (or Texas and Arizona) combined, and the melting trend is accelerating. Sea-ice extent in summer has declined more dramatically than the annual average, with a loss of 15-20% of late-summer ice coverage. Additional declines of 10-50% in annual average sea-ice extent are projected by 2100. Loss of sea-ice during summer is projected to be considerably greater, with a 5-model average projecting more than a 50% decline by late this century, and some models showing near-complete disappearance of summer sea-ice.

Average air temperatures across the Arctic Ocean from January to August 2006 were about 2-7 degrees F warmer than the long-term average across the region during the preceding 50 years. The National Snow and Ice Data Center

---

[4] The ACIA is a "comprehensively researched, fully referenced, and independently reviewed evaluation of arctic climate change and its impacts for the region and for the world. It has involved an international effort by hundreds of scientists over four years, and also includes the special knowledge of indigenous people."

reported that the amount of sea-ice in 2007 was the lowest on record and the pace of melting was accelerated. The onset of freeze-up was later than normal and the onset of melt happened earlier. These changes appear to be driven largely by a positive feedback loop called the albedo effect. The albedo effect involves reduction of the extent of lighter-colored sea-ice or snow, which reflects solar energy back into the atmosphere. Consequently, there is a corresponding increase in the extent of darker-colored water or land that absorbs more of the sun's energy. This greater absorption of energy causes faster melting, which in turn causes more warming and creates a self-reinforcing cycle that becomes amplified and accelerates with time. Scientists believe that this cycle caused a tipping point in Arctic sea-ice thinning in the late 1980s, sustaining a continual decline in sea-ice cover that cannot be reversed easily. Earlier melt onset and lengthening of the melt season result in decreased total ice cover at summer's end. Climate models now project that abrupt reductions in the extent of summer ice are likely to occur over the next few decades and that near ice-free September conditions may occur as early as 2040 (Holland et al., 2006).

The projected temperature increases and ice loss in the polar regions are also expected to alter ocean currents and salinity patterns. This may force the displacement of some marine species into areas of higher salinity, i.e., where more truly "oceanic" conditions prevail. Some scientists have warned that global warming in excess of 1 degree C could trigger a runaway melting of the world's ice sheets. They also warn that air pollution, particularly soot, may quicken the ice melting by decreasing the reflectance, and increasing the absorbance, of heat radiation by water crystals. Even a few parts per billion of soot in snow can cause significant effects.

In the Arctic, higher ocean temperatures and lower salinities, contraction of seasonal sea-ice extent, rising sea levels, and a host of other effects are certain to have significant impacts on marine species. Sea-ice is a highly dynamic habitat with different types, forms, stages and distributions that all operate as a complex matrix in determining biological productivity and marine organisms use. For marine mammals adapted to life with sea-ice, the effects of reduction in ice are likely to be reflected initially by shifts in range and abundance. Demographic changes associated with shifts in geographic range will likely be observed as decreased recruitment in areas of reduced sea-ice. The challenge for species to accommodate such change is increased because of the large area involved, the rapid rate at which the warming is predicted to occur, large inter-annual variation in climate, and the accelerated pace of human development.

Climate change will have substantial and possibly irreversible consequences on sea-ice and the ice-dependent marine mammals. The most serious threats to Arctic marine mammals are the loss of sea-ice habitat and the unique ecosystem with which it is associated, and the related increasing human activities that result from easier access to the region.

SOUTHERN OCEAN AND THE ANTARCTIC

Similar to the Arctic, Antarctica and the Southern Ocean have seen some of the most dramatic effects of global warming. However, there are important differences in the type of sea ice found in Antarctica. In contrast to the Arctic, most of the sea ice around Antarctica is less than 1 year old, waxing and waning on an annual cycle (Nicol et al., 2008). At minimum extent in February, sea ice covers an area of approximately 4 million km^2 of the Southern Ocean (Gloersen et al. 1992). However, at maximum extent in September the sea ice cover expands to approximately 19 million km^2, an area 1.5 times that of the Antarctic continent (13.2 million km^2), and slightly greater than the area of South America (17.8 million km^2). This change from open ocean to ice cover represents one of the greatest seasonal changes in physical properties anywhere on earth and has an extraordinary influence on oceanic and atmospheric circulation and marine ecosystems.

Air temperatures in the Antarctic Peninsula region have risen by more than 2°C in the past 50 years, about 5 times faster than the global mean rate. The largest regional warming on Earth in the middle troposphere - the layer of the atmosphere around 5 km above ground - is occurring over the Antarctic (Tuner, 2006). The Western Antarctic Peninsula has warmed more than four times faster than the average rate of Earth's overall warming over the last fifty years. (IPCC, 2007a). The Southern Ocean has warmed to 3,000 m deep (Jacobs, 2006). There has been an overall reduction in total ice shelf area on the Antarctic Peninsula by 27,000 km2 in the last 50 years[5] (ASOC 2009). While the location and rate of warming in Antarctica has not been uniform, areas of significant warming have exhibited profound changes on land and in

[5] Some ice shelves, such as the Wordie (>1000 km2), Larsen (>6500 km2), Prince Gustav (2000 km2) and the Wilkins (14000 km2) Ice Shelves underwent rapid and sometimes catastrophic disintegrations (ASOC 2009).

Recent observations have shown that the Amundsen Sea sector – holding approximately one third of the ice of the entire West Antarctic Ice Sheet - is currently the most rapidly changing region of the entire Antarctic.

the ocean (IPCC, 2007b). Warmer temperatures have melted sea ice habitat which has caused reductions in populations of Antarctic krill, Weddell seal and Adélie and emperor penguin. In contrast, species such as shallow-water sponges and chinstrap and gentoo penguins, that do not like sea ice, have expanded into ice-free territories (Ainley et al., 2005; Atkinson et al., 2004; Ducklow et al., 2007).

At the current warming rate, climate models predict that the average global temperature could reach 2°C above preindustrial levels by 2042 (Tynan and Russell, 2008). The models project that the ocean surface would warm by more than 0.5°C with greater increases downstream of Australia (Tynan and Russell, 2008). With 2°C temperature increase, the Southern Ocean's sea ice coverage will shrink an average of 10-15% and up to 30% in some areas (Tynan and Russell, 2008). As such ice-dependent species, such as the Antarctic minke whale are projected to lose between 5-30% of ice-associated habitat within 40 years (Tynan and Russell, 2008). The implications of this habitat loss are particularly troubling considering this will occur in roughly an individual whale's life time.

A four-year study of observations of atmospheric CO_2 from 40 stations around the world reveals that the Southern Ocean is releasing CO_2 into the atmosphere because it is saturated due to climate change (Le Quéré et. al. 2007). Greenhouse gases and ozone depletion has caused an increase in winds over the Southern Ocean which has released stored CO_2 and is preventing further absorption of the greenhouse gas (Le Quéré et. al. 2007). Essentially, the world's oceans are losing their ability to absorb carbon dioxide (CO_2) because they are already so full of CO_2. Since 1981, the Southern Ocean sink stopped increasing, while CO_2 emissions increased by 40% over the same period. In 2006, human activity released 9.9 billion metric tons (2.18×10^{13} pounds) of carbon, an increase from 8.4 billion metric tons (1.85×10^{13} pounds) in 2000 (Canadell et. al., 2007). Atmospheric concentrations of CO_2 reached 381 parts-per-million in 2006 after CO_2 emissions from burning fossil fuels rose to 8.4 billion metric tons (1.85×10^{13} pounds) annually, according to figures from the United Nations, British Petroleum and the U.S. Geological Survey (Canadell et. al., 2007). Oceans and plant growth absorbed only around 540 kilograms per metric ton (1,190 pounds per short ton) of the CO_2 produced in 2006, compared with 600 kilograms per metric ton (1,322 pounds per short ton) in 2000. Coupled with an emissions growth rate of 3.3 percent—triple the growth rate of the 1990s—the atmospheric burden is now rising by nearly two parts per million of CO_2 a year, the fastest growth rate since 1850, the international team of researchers reports in Proceedings of the National Academy of Sciences USA.

The Earth's carbon sinks - of which the Southern Ocean accounts for 15% - absorb about half of all human carbon emissions (Le Quéré et. al. 2007). During this time period, shifts toward the pole of westerly winds in the Southern Ocean reduced the region's ability to absorb CO_2 by 30% (Le Quéré et. al. 2007). The ocean is absorbing the same quantity of the gas as it was 24 years ago which contradicted scientists' expectations that the amount of CO_2 absorbed would increase proportionate to rising levels in the atmosphere (Le Quéré et. al. 2007). However, increased average wind speeds over the Southern Ocean linked to climate change and the depleted ozone layer have impacted the ocean's ability to absorb more gas (Le Quéré et. al. 2007). Because the world is warming unevenly, pressure differences between different regions are increasing. One consequence of this is increasing wind speed in the Southern Ocean. The ozone layer's depletion has also contributed to higher winds. Strong winds mix and circulate more water which brings up deep colder water saturated with CO_2 so that it cannot accept any more from the atmosphere.

More CO_2 will remain in our atmosphere with the Southern Ocean reaching its saturation point. This new research suggests that stabilization of atmospheric CO_2 is even more difficult to achieve than previously thought. Additionally, acidification in the Southern Ocean is likely to reach dangerous levels earlier than the projected date of 2050.

OCEAN ACIDIFICATION

The oceans are an important place for the exchange of carbon dioxide - carbon dioxide in the atmosphere dissolves in the oceans and the oceans can release it back into the atmosphere (Raven et al., 2005). The oceans' biological absorption is almost as great as terrestrial absorption due to the much greater surface area of the oceans despite the fact that terrestrial systems have a higher biological uptake of CO_2 per unit area (Field et al., 1998).

"Ocean acidification" refers to the continuous decrease in the oceans' pH level, caused by their absorption of human produced atmospheric carbon dioxide (Bass and Simmonds, 2006). CO_2 in the atmosphere is relatively inert, but in ocean water it reacts chemically, biologically and geologically, affecting many processes in marine biogeochemistry. Pure ocean water is naturally alkaline, having a pH between 8 and 8.3 (Raven et al., 2005). When CO_2

dissolves in sea water, it reacts to form ionic and non-ionic chemicals (H_2CO_3 (carbonic acid) + HCO_3^- and CO_3^{-2} (bicarbonate and carbonate ions) + H^+ + $CO_{2\,(aq)}$). When the hydrogen ion (H^+) increases, it reduces the oceanic pH and the sea water becomes more acidic.

This acidity impacts organisms that need carbonate ions to build calcium carbonate ($CaCO^3$) shells (Doney, 2006). Put simply, H^+ ions combine with carbonate ions, making bicarbonate ions and decreasing the carbonate ions ($CaCO_3^{2-}$). This is significant because carbonate ions are necessary to prevent calcium carbonate from dissolving in sea water and therefore be available to shell builders to make CaCO3 structures (Raven et al., 2005).

Oceanic carbonate ion concentration is expected to drop by half over this century (Bass and Simmonds, 2006; Doney, 2006). Some oceanic plankton species are highly sensitive to changes in CO_2 concentrations in sea water. The calcification rate of all calcifying organisms investigated to date decreased in response to a decreased calcium carbonate saturation state (Feely et al., 2004; Raven et al., 2005). Calcifying organisms that may be affected include the coccolithophores,[6] pteropods,[7] gastropods and foraminifera which are major food sources for fish and some whale species (Riebesell et al., 2000; Doney, 2006). Regional variations in pH will mean that by 2100 the process of calcification may become extremely difficult for some of these groups of organisms, particularly in the Southern Ocean (Raven et al., 2005; Orr et al, 2005).

Different water levels have had carbonate ions levels associated with them. Cold high latitude waters are naturally less supersaturated with respect to carbonate ions than warm waters and do not contain calcifying organisms because their shells would dissolve. Surface waters are supersaturated with carbonate ions (CO_3^{2-}) and do not dissolve calcifying organisms (Doney, 2006). As more atmospheric CO_2 dissolves in surface waters, shallower depths will be under-saturated with carbonate ions, and therefore capable of dissolving calcium carbonate structures (Orr et al., 2005; Doney, 2006).

High-latitude oceans will be the first to show measurable impacts of ocean acidification, including reduced carbonate availability and decreased habitat suitable for calcifying plankton due to rising saturation levels within the water

---

[6] Increased CO2 (lowered pH) also results in reduced calcite production in cultures of two dominant marine calcifying species, the coccolothophorids Emiliania huxleyi and Gephryrocapsa oceanica. These single cell microalgae are important primary producers and an important part of marine ecosystems. E. Huxley is one of the most prominent producers of calcium carbonate in temperate and sub-polar oceans. (Riebesell et al., 2000).

[7] Research has found that pteropods experience reduced calcification when exposed to elevated CO2 (Orr et al., 2005).

column (Doney, 2006). Pteropods are prominent components of the upper-ocean biota in the Arctic Ocean, Southern Ocean and subarctic Pacific Ocean and are the major planktonic producers of aragonite which is one of the main structures of calcium carbonate (CaCO3) (Orr et al., 2005). In the Antarctic marine food web, pteropods are an important food source for predators and in some regions of the Southern Ocean pteropods are the dominant zooplankton group (Cabal et al., 2002; Seibel and Dierssen, 2003). The pteropod *Limacina helicina* can make up more than one-fifth of the zooplankton biomass in McMurdo Sound (Siebel and Dierssen, 2003) and is an important prey item for many Antarctic species, including whales and fish (Foster and Montgomery, 1993; Davis et al., 1999).

If current CO_2 emissions continue, it is predicted that ocean acidification will be severe enough to dissolve pteropod shells within 50 years and polar pteropods could disappear completely by 2100 (Seibel and Fabry, 2003; Orr et al., 2005). This could have catastrophic impacts on the entire Southern Ocean food chain, affecting predators which exclusively feed on pteropods such as the zooplankton *gymnosome* (Seibel, & Dierssen, 2003) as well as the diets of many species of zooplankton, North-Pacific salmon, mackerel, herring, cod and baleen whales of which pteropods are a key primary component (Bass and Simmonds, 2006; Lalli and Gilmer, 1989; Seibel and Dierssen, 2003; Orr et al., 2005).

The potential food web impacts are significant. These range from changes at the lowest levels where non-calcifying organisms may fill the space and change ecosystem structures and processes (Bass and Simmonds, 2006; Riebesell et al., 2000; Turley et al., 2004; Raven et al., 2005). Multi-cellular organisms could experience decreased reproductive potential, slower growth or increased disease susceptibility (Ishimatsu et al., 2004; Raven et al., 2005). Squid, which are key prey for many deep-diving marine mammal species, such as the beaked and sperm whales, may be especially vulnerable to ocean acidification (Rosa and Seibel 2008). Their high-energy swimming method and high metabolism require a good supply of oxygen, and their capacity to carry oxygen is reduced by lowered blood pH (TRS 2005; Rosa and Siebel 2008). These impacts can cascade through food webs and affect ecosystem structure (Raven et al., 2005).

CLIMATE CHANGE IMPACTS TO CETACEANS

Depending on the global temperature rise, IPCC estimated that between 20 and 70 percent of all plant and animals species assessed so far face an increased risk of extinction[8] (IPCC, 2007b). However, the complexity of oceanographic factors – water temperature, sea-ice coverage, sea level, salinity, acidity, currents – along with the complexity of marine food webs, prey availability, and climate patterns such as wind speed, storm frequency, rainfall, make predicting oceanic changes a challenge (Simmonds and Isaac 2007).

More challenging is trying to predict the impacts on ocean-dwelling creatures, including whales and their prey. There are so many gaps in information about marine species, including whales and what they eat. Basic information about many species including abundance estimates, population trends, habitat and prey needs is unknown. The IUCN red list classifies 60% of the 67 cetacean species as "data deficient," meaning future research needs to be a priority (IUCN 2008). With better information, more species could well prove to be in danger. At least a quarter of the world's cetaceans are considered endangered and the situation may be worse as the status of many others remains unclear. The blue whale (*Balaenoptera musculus*), fin whale (*Balaenoptera physalus*) and sei whale (*Balaenoptera borealis*) all remain listed as Endangered, pending more evidence of recovery[9] (IUCN 2008). Climate change threats add pressure to already vulnerable species.

---

[8] If global temperatures rise by more than 1.5 - 2.5°C, IPCC projects that between 20-30 percent of plant and animals species assessed so far are likely to be at increased risk of extinction (IPPC, 2007b). If the global temperature rise is greater than 3.5°C, 40-70% of assessed species are at risk of extinction (IPCC, 2007).

[9] The Cetacean Specialist Group of the IUCN Species Survival Commission assessed the conservation status of most cetacean species for the 2008 IUCN Cetacean Red List. Almost a third of cetaceans changed their Red List status, with the majority being at greater risk than previously (IUCN 2008). The 2008 Cetacean Red List is a joint product of IUCN and the Global Mammal Assessment. The IUCN Red List threat categories are the following, in descending order of threat:

- Extinct or Extinct in the Wild;
- Critically Endangered, Endangered and Vulnerable: species threatened with global extinction;
- Near Threatened: species close to the threatened thresholds or that would be threatened without ongoing specific conservation measures;
- Least Concern: species evaluated with a low risk of extinction;
- Data Deficient: no evaluation because of insufficient data.

Critically Endangered (Possibly Extinct) is not a new Red List category, but is a flag developed to identify those Critically Endangered species that are in all probability already Extinct but

The numerous existing impacts on cetaceans such as ship strikes, fisheries bycatch, oil and gas development activities, unsustainable hunting, chemical and noise pollution further confound the predictions about how climate change will affect whales. Nobody knows whether different whale species' can adapt to oceans changed by global warming, particularly if changes continue to occur more rapidly than predicted and existing and new human impacts continue and increase.

Some cetacean species and populations are likely to be impacted by climate changes. It is unknown exactly what impacts climate change will have on whales, but it is likely to impact several factors, including:

- habitat availability
- ocean temperature
- changes in sea-ice distribution
- prey availability

These factors will likely impact feeding, breeding and the ability to survive. Usually these factors are interrelated, meaning impacts to one may strongly affect one or more of the other factors. For example, preferred habitat may exist because of a satisfactory ocean temperature, amount of sea ice and related prey availability. Upsetting one factor affects the others. This is particularly true for species for which sea ice provides an important habitat and/or for their prey.

HABITAT LOSS

Researchers believe that Arctic marine mammals may experience multiple negative impacts from climate change, the majority of which involve profound changes in habitat (Ragen 2008). Habitat loss has been "the primary factor responsible for the rapid rate of species extinctions and the global decline in biodiversity in the past one hundred years" (Dayton et al. 2002). Climate change has the potential to degrade habitat more than any other human-caused factor (Thomas et al. 2004). However, it is hard to predict the extent of climate-related habitat degradation; relationships between climate and habitat features are not well understood, human contributions to climate change

for which confirmation is required (for example, through more extensive surveys being carried out and failing to find any individuals).

continue at significant levels, and various other factors contribute to habitat degradation (Ragen 2008). Thus, it is challenging to discern the relative effect of climate change from the effects of other risk factors (Ragen 2008). Climate models predict that the physical manifestations of climate change will continue to progress (ACIA 2004), but it is not clear exactly what the direct and indirect biological and ecological consequences of these changes will be. However, evidence of effects is increasing (Stirling and Smith 2004, Stirling et al. 2004) as well as anomalous events that have not been seen before by researchers who have long studied Arctic marine mammals such as cannibalism, starvation, drowning, stranding and stampeding.

Important habitat for coastal whale species and species that require coastal bays and lagoons for breeding, such as grey whales (*Eschrichtius robustus*) and humpback whales (*Megaptera novaeanglieae)*, could be adversely affected. It is possible that the Arctic Ocean and its bays could become more productive due to warming, benefiting some top predators. However several cetacean species that are found year-round in the Arctic depend on the productive sea ice edge in varying degrees for foraging as well as use ice cover to avoid predation. These ice-associated cetaceans are likely to be negatively impacted by climate change induced sea ice reductions.

It is unknown if ice-association whale species in the Arctic can adapt to loss of sea ice and the associated changes of prey distribution and availability. Bowhead whales (*Balaena mysticetus*) typically select habitats with moderate to light ice cover (Moore et al 2000). Beluga whales (*Delphinapterus leucas*) are consistently associated with heavy ice cover, possibly reflecting their preference for feeding on ice-associated arctic cod (Moore et. al. 2000). Narwhals (*Monodon monoceros*) are so well adapted to life in very dense ice that their Baffin Bay wintering habitat can become so thick with ice that there is less than .5% of open water (Laidre and Heide-Jørgensen 2005). Researchers posit that narwhals return annually to ice-dense areas to find predictable prey as they have observed intense feeding activity (Laidre and Heide-Jørgensen 2005).

Reduced ice cover may reduce foraging opportunities for beluga whales but could expand them for bowhead whales (Moore 2000). It is unknown, however, whether the bowhead whale would be able to adjust to open water without ice (Tynan and DeMaster 1997) and this species may not be able to tolerate increased water and air temperatures (IWC 1997; Bannister 2002). Researchers have identified significant reductions in sea ice cover in half of the sixteen assessed habitat regions of seasonal importance to bowhead whales (Moore, Laidre. and Overland 2006).

A warmer Arctic with less ice and more open water may cause more temperate species to expand their normal distributions northward. This could result in food web changes and increase competition between species for space and prey. Also, species that currently migrate into Arctic waters to feed may be adversely affected if the prey they seek are reduced or have moved outside of traditional feeding grounds (Simmonds and Isaac 2007). Finally, species that move north may not be able to find suitable habitat.

TEMPERATURE INCREASE

Temperature is an important factor in cetacean distribution and species have evolved to live within certain temperature regimes (Simmonds & Eliott 2009). As ocean temperatures increase, some species will shift distributions and move to the water temperatures they prefer. For some species and populations, however, such movement will not be possible, particularly for species with restricted geographical distributions that cannot expand their ranges.10 Climate change is likely to decrease the range of all the cetacean species that IUCN has listed as threatened for which predictions can be made (Simmonds & Elliott 2009). Over their evolutionary history, many marine mammal species have adapted their behaviors and distributions in response to changing environmental conditions. However, it is unclear whether cetaceans will be able to adapt to the relatively rapid rate of climate change which is predicted (Simmonds and Isaac, 2007).

Temperature increases will affect not only cetaceans, but the species upon which they prey. Changes in availability, locality and abundance of food resources, especially for species with specialized feeding habitats, could prove problematic (Simmonds and Elliott, 2009). Long-term plankton monitoring in the North Atlantic has demonstrated recent changes; warmer water species are currently increasing in the North Sea due to regional climate warming and the effects of the North Atlantic Oscillation (Edwards et al., 2007). The warmer

---

[10] The endangered vaquita (Phocoena sinus) is limited to the warm waters at the northern end of the Gulf of California in a closed embayment and cannot move northward to find cooler water or alternative prey if temperatures rise and prey is not available (Simmonds and Elliott 2009). The Ganges river dolphin (Platanista gangetyica) and the boto (Inia geoffrensis) may be vulnerable to temperature changes within their restricted riverine habitats (Simmonds and Elliott 2009; Learmonth et al., 2006; Simmonds and Isaac, 2007). The Chinese or Yangtze river dolphin (Lipotes vexillifer) has very restricted habitat and IUCN listed the species as Critically Endangered (Possibly Extinct) (Simmonds and Elliott 2009).

water species are not replacing the colder water species in similar abundances which affects other trophic levels including fish larvae (Simmonds & Elliott, 2009). For example, an important zooplankton species has declined by 70% in the North Sea (Simmonds & Elliott, 2009).

In the same region, it appears that cetacean distributions have changed. There has been a decline in relative frequencies of strandings and sightings of white-beaked dolphins, (Lagenorhynchus albirostris), a colder-water species, and a relative increase in strandings and sightings of the common dolphins, (Delphinus delphis) a warmer water species (MacLeod et al., 2005). This suggests a possible range contraction of white-beaked dolphins and a range expansion of common dolphins due to temperature change or competition (Simmonds and Elliott, 2009). White-beaked dolphins may be particularly adversely impacted because they generally inhabit shelf waters around northwest Europe and there is not similar suitable habitat further north (MacLeod et al., 2005).

Prey Availability

Different whale species that migrate to or occupy polar waters rely on sea ice habitat in varying ways. In the Southern Ocean, Antarctic minke, blue, humpback, sperm and fin whales migrate to the rich ice frontal zones where they forage (Tynan 1997, Tynan 1998).The bowhead whale in the Arctic prefers pack ice with leads of open water, although it can travel below ice. The North Atlantic narwhal occupies very densely ice-filled waters in the winter (Laidre and Heide-Jørgensen 2005). These whales rely on sea ice for many reasons, but perhaps most importantly for its connection to prey availability.

In icy polar waters, the ice is integral to prey production of the smallest species. Krill (*Euphausia superba*), lives in the open ocean and is the foundation prey species for many high latitude animals. Krill occurs mainly in large swarms and reaches particularly high numbers in Antarctica. Antarctic krill can grow up to a length of 6 cm and can live for 5-6 years. They are one of the largest protein resources on Earth and are eaten by fish, seals, penguins and whales.

To a large extent, the krill-based ecosystem is spatially coincident with the sea ice zone and the krill life-cycle has evolved in association with sea-ice and regional circulation patterns (Nicol et al., 2008). Larval krill have an overwintering strategy that includes feeding on ice algae scraped from the underside of annual ice (Daly, 1990). Thus, krill occur in regions that are covered by ice in winter, and their life cycle is closely related to the oscillation of sea ice cover (Brierley and Thomas, 2002). Krill also graze on ice algae at the ice edge in the marginal ice zone (Daly and Macaulay, 1991). These linkages between krill and sea ice suggest that a reduction in ice-habitat would co-occur with a reduction in krill production (Nicol et al., 2008).

The Antarctic krill population appears to have dropped by about 80% since the 1970s, most likely declining due to the dramatic decline in winter sea-ice (Tynan and Russell 2008). The Antarctic Peninsula has warmed by 2.5°C in the last 50 years, with a striking decrease in sea-ice. Also in Antarctica, frontal zones are where water masses of different temperatures meet. These productive areas are associated with upwelling of nutrients that support large plankton populations on which species such as Antarctic krill feed (Tynan and Russell 2008). As frontal zones move southward because of contracting sea ice, they move closer together, reducing the overall area of foraging habitat available. As the krill is dependent on sea ice, less sea ice is also expected to reduce the abundance of food for whales in the feeding areas (Tynan and Russell 2008).

As seasonal visitors to the Southern Ocean, it is hypothesized that baleen whales may spend between 90 and 120 days foraging in Antarctic waters (although there is scant empirical data to support this hypothesis) (Nicol et. al., 2008). Baleen whales efficiently exploit krill because of their tendency to be found in high densities and in predictable locales (Nicol et. al., 2008). Five species of krill are common in the Southern Ocean - *Euphausia superba* (Antarctic krill), *Euphausia crystallorophias* (ice or coastal krill), *Euphausia frigida*, *Euphausia triacantha* and *Thysanoessa macrura* (Nicol et. al., 2008). Of these, Antarctic krill is the most abundant and occurs in very large aggregations, and is the major component of most baleen whale diets, despite a large variety of prey items being recorded (Nicol et. al., 2008). Blue whales and minke whales are generally regarded as Antarctic krill specialists (Kawamura 1994), but both species consume other krill species on occasion. Both species are thought to be most associated with sea ice and continental shelf and shelf break areas (Kasamatsu et al., 2000; Murase et al., 2002) and occur mainly in southernmost latitudes, usually aggregated close to the edge of the pack ice (Kasamatsu et al., 1998; Branch et al., 2007) and in the case of minkes sometimes within the pack ice itself.

In the Arctic, sea ice appears to boost the productivity in the Bering, Chukchi and Beaufort Seas. Scientists believe areas that freeze in the winter experience an algae bloom early in the spring related to the plants that flourish under the ice. The frigid water that time of year limits the number of tiny, algae-eating zooplankton. That means much of the algae falls to the bottom when it dies, a feast for bottom-dwellers and bottom-feeders. Where ice has vanished, algae do not bloom until later in the spring when the water has warmed slightly and zooplankton are abundant. The zooplankton consume the algae, so less food falls to the bottom and less algae is available to bottom-feeders. Instead, the zooplankton nourish a food chain that favors pelagic fish and fish-eating mammals. Thus, the loss of sea ice may cause community restructuring. Scientists have already measured a decrease in nutrients on the bottom and believe that once the ecosystem is altered due to warming temperatures and loss of sea ice, it will be very difficult to regain the productivity of the sea ice food web (Tynan and DeMaster 1997).

The bowhead whale's foraging efficiency is intricately linked to the Arctic ecosystem by changes in ice cover, in spring ice break-up, in algal blooms, and in the abundance of its prey species. If plankton species are affected by climate change, this could lead to cascading effects through the food chain. In addition, global warming and possible shifts in wind patterns could also affect the distribution of polynyas in the polar ice cap. Dark polynyas often contain significant blooms of phytoplankton. Cetacean species such as bowhead whales that rely on ice edges for phytoplankton foraging might be adversely affected by any decline in these habitat areas.

Year to year variability in seasonal distributions of bowheads in the Beaufort and Chukchi Seas may be influenced by ice cover and feeding opportunities (Moore and Reeves, 1993). Off Alaska, bowhead whales feed primarily on zooplankton, especially large copepods (such as *Calanus glacialis* and *C. hyperboreus*) and euphausiids (Lowry, 1993). It follows that the bowheads' ability to find suitable and reliable amounts of zooplankton in a warmer Arctic will determine their movements and redistribution (Tynan and DeMaster 1997).

Also reliant on the undersea ice algae is the arctic cod, a major prey species for beluga whales and narwhals (Tynan and DeMaster 1997). Arctic cod exceeds all other prey items in arctic regions in abundance and energetic value (Finley et al., 1990a). The distribution and diet of arctic cod vary with ice conditions and large numbers of fish can occur in areas of marginal ice zones (Tynan and DeMaster 1997). Therefore, the trophic cascade upon which arctic marine mammals depend is ultimately driven by the complex interactions between ice, light penetration, nutrient supply, and productivity (Tynan and DeMaster 1997). Cryopelagic fishes such as arctic cod (*B. saida*) are especially adapted to feed under the ice where they rely on crustaceans associated with the ice undersurface and the adjacent water column (Lønne and Gulliksen, 1989), primarily amphipods (i.e., *Onisimus* spp., *Apherusa glacialis*, and *Parathemisto libellula*) and copepods (i.e., *Calanus glacialis* and *C. finmarchicus*) (Alexander, 1981; Dunbar, 1981; Bradstreet et al., 1986; Lønne and Gulliksen, 1989; Crawford and Jorgenson, 1993). Behaviors of arctic cod that lead to the formation of large aggregations are probably quite crucial to the foraging of higher vertebrates (Tynan and DeMaster, 1997). Otherwise, as noted by Welch et al., "It is difficult to imagine such large whale concentrations finding maintenance rations in the form of dispersed arctic cod…." Also difficult, is predicting how arctic cod may be redistributed in a warmer Arctic (Tynan and DeMaster, 1997). However, since their life history is closely linked to sea ice, regional sea ice extent changes may lead to redistributions of arctic cod, and consequently to redistributions and altered migrational patterns of marine mammals (Tynan and DeMaster, 1997).

Beluga whales and narwhal migrations are linked to the spring production of ice algae and ice-edge productivity. These species migrate long distances, leaving wintering grounds further south to inhabit the Canadian archipelago during the open-water season (Davis et al., 1980). Feeding aggregations of these three species in the Canadian High Arctic occur primarily in nearshore areas, where dense schools of arctic cod concentrate in late summer (Bradstreet et al., 1986). Beluga whales and narwhals often congregate at the ice edge and have been observed to dive repeatedly under the ice, presumably to feed (Bradstreet, 1982; Lydersen et al., 1991). However, coastal feeding aggregations may occur in open water, as well as beneath or near ice.

ARCTIC WHALES

The Arctic marine environment is considered to be the region where sea ice is a dominant feature for a considerable part of the year, including the Arctic Ocean, the Bering and Barents seas and Hudson and Baffin bays (Huntington and Moore, 2008). Many species visit the Arctic at various times or are found near sea ice at some point in their migrations or life cycle. Sea ice appears to be preferred habitat for many species, although relatively few actually require sea ice for their survival. Some species are found year-round in the Arctic but also have separate stocks in more southerly locations. Three cetacean species occupy the Arctic environment and at least some portion of the population is associated with sea ice year-round:

- Bowhead whale (Balaena mysticetus)
- Beluga whale (Delphinapterus leucas)
- Narwhal (Monodon monoceros) (Huntington and Moore, 2008).

Five additional cetacean species are seasonal or occasional migrants to the Arctic or have some association with sea ice:

- Gray whale (Eschrichtius robustus)
- Killer whale (Orcinus orca)
- Minke whale (Balaenoptera acutorostrata)
- Fin whale (Balaenoptera physalus)
- Humpback whale (Megaptera novaeangliae).(Huntington and Moore, 2008).

Laidre et al. (2008) attempt to quantify the sensitivity of Arctic marine mammals to climate change and suggest an index of three sensitivity dimensions based on:

1. the narrowness of distribution and specialization in feeding
2. seasonal dependence on ice
3. sea-ice as a structure for access to prey and predator avoidance.

They conclude that the narwhal appears to be the most sensitive cetacean species because of its reliance on sea-ice, their specialized feeding and their winter habitat within the ice fields (Laidre et al., 2008). They ranked the bowhead in the middle because it migrates away from the ice in the winter, and the beluga as the least sensitive cetacean because it is widely distributed (Laidre et al., 2008). The five whale species that seasonally occupy the Arctic and sub-Arctic habitats may move further north, stay longer and compete with the species which have been year-round Arctic occupants (Moore and Huntington, 2008).

Responses in ice-associated species populations will depend on the dynamics of trophic cascades associated with annual cycles of production in sea ice-associated communities (e.g., Arrigo and van Dijken, 2004, Bluhm and Gradinger, 2008). While sea ice reductions will likely have negative effects on seals that birth pups on ice (i.e., harp, hooded, ribbon, and spotted seals), the reliance of ice-associated whales (beluga, narwhal, and bowhead) on sea ice-mediated ecosystems is unclear (Laidre et al., 2008). While these cetaceans are highly adapted to the Arctic's icy waters, they can also survive tens to hundreds of kilometers from sea ice, and sometimes occupy open-water for part of the year (e.g., Moore, DeMaster and Dayton, 2000). Sea ice reductions may actually boost feeding opportunities for bowhead whales and their prey in their summer and autumn habitats (Moore and Laidre, 2006). During roughly two decades of sea ice loss in the Alaskan Beaufort Sea, the western Arctic population has increased at 3.4% (George et al., 2004; Walsh, 2008); this growth rate suggests current trends in sea ice reduction are not impeding recruitment to this population as it recovers from commercial whaling overhunting. Trends in population size estimates for narwhals and belugas do not exist so no similar comparisons can be made.

Narwhals

The narwhal is perhaps the world's most unique looking cetacean, with the males growing a spiraled tusk that can reach nine feet in length. It also has the distinction of being the Arctic's most specialized, range-restricted and northernmost cetacean, believed to number between 50,000 and 80,000 in the world (Laidre and Heide-Jørgensen, 2005). Eighty to ninety percent of the world population inhabits Canada's Baffin Bay, wintering in the deep water of central Baffin Bay and summering thousands of miles away in shallow bays and fjords of the High Arctic. A second group of around 5,000 occupies the northern part of Canada's Hudson Bay and an even smaller population of unknown number lives east of Greenland.

Winter is a busy time of mating and feeding for the narwhal, despite the extreme cold which can reach below minus 30 degrees F and thick ice cover (Laidre and Heide-Jørgensen, 2005). On a typical winter day, narwhals dive non-stop down to 5,000 feet, grazing on the seafloor and gorging on fatty energy-rich Greenland halibut, known as turbot (Laidre and Heide-Jørgensen, NOAA website). The Baffin Bay population needs to consume approximately 880 tons of turbot daily during the winter, and during the milder summer months, they eat little. At the deep diving depths, narwhals are sustained solely by their highly oxygenated blood and muscles because the deep-sea pressure collapses their lungs. Dives can last over 30 minutes and when narwhals surface, they must find small, sometimes fleeting, cracks in thick winter pack ice to breathe. Open water can freeze suddenly during windless conditions and sub-zero temperatures. In the spring, the pack ice loosens, signaling that it is time for the narwhals to start its two-month migration north.

In 2008, researchers ranked narwhals as the most vulnerable Arctic mammal to climate-induced habitat change (Laidre and Heide-Jørgensen, 2008). This is due to the narwhal's small range, narrow migration routes, limited world population, restricted diet and low genetic diversity (Laidre and Heide-Jørgensen, 2008). It is unclear exactly what the climate changing impacts will be and the suggested scenarios are very different.

One counterintuitive possibility is that there will actually be cooling in that region, reducing the scant open water that ensures winter survival. Global warming can cause increased rainfall in coastal regions; this increases freshwater and decreases ocean salinity, slowing or shutting down ocean currents that would normally deliver warmth from the Equator. Since 1978, Baffin Bay sea ice cover has been increasing steadily and the percentage of open water at winter's end had shrunk to an average of just ½ a percent (Laidre and Heide-Jorgensen, 2005). Dense or complete ice cover may act as a barrier to cetaceans as they need to breathe at the surface (Lowry, 2000). When sufficient open water is not available, large-scale mortality events may occur. Ice entrapments are best documented for narwhals and belugas, where sudden changes in weather conditions cause rapid freeze-up of leads and cracks, thus eliminating access to oxygen (Siegstad and Heide-Jørgensen, 1994, Heide-Jørgensen et al., 2002).

During the last four years, however, that seems to have reversed and there is less ice cover. (NWF Laidre) Increased open water is also a concern for two reasons. First, rising water temperatures could make the ecosystem less productive, adversely affecting the narwhal's primary prey source of cold-water turbot. Second, commercial fishing will become more desirable in the narwhal's previously ice-locked feeding areas.

Currently, the narwhals short-term prospects appear positive. The populations seem stable in the decade since the Canadian and Greenland governments implemented hunting quotas with the region's Inuit peoples. The Inuit take several hundred narwhals annually, relying on the nutritious meat and skin and the valuable male's 9 foot long tusk.

Exactly how and when climate change will impact narwhals remains to be seen.

CHANGES IN DISTRIBUTION

Scientists expect that many sub-arctic marine mammal species will extend their range into the Arctic as temperatures and habitats change. These species may compete with Arctic marine mammals for prey or habitat, alter predatory-prey relationships, and introduce diseases and parasites new to Arctic marine mammals.

If sea ice reduction trends continue, seasonally migrant cetaceans will likely range farther north and stay longer, including fin, humpback, minke, gray, and killer whales11 (Moore and Huntington, 2008). These species now occur near or within seasonal ice in the Bering Sea and Barents Sea and may follow the forage fish stocks north if they increase as predicted to accompany sea ice reductions (Hunt et al., 2002; Øien, 2003; Bluhm and Gradinger, 2008). Indeed, in the southeastern Bering Sea where dramatic sea ice reductions have occurred (Overland and Stabeno, 2004), fin whale calls have been recorded throughout winter (Moore and Huntington, 2008), suggesting that at least some of the whales that occupy the Bering Sea during summer (Moore et al., 2000, Moore et al., 2002, Tynan, 2004) remain there during the winter. In addition, at least some humpback whales remain in southeast Alaska and offshore Kodiak, Alaska, through winter, seemingly in response to the availability of herring (Moore and Huntington, 2008). Such flexible behavior indicates a species that can adapt their migration habits based upon opportunity (Moore and Huntington, 2008).

GRAY WHALES

Annually, gray whales travel 12,000-miles round-trip from the cold Bering Sea to the warm waters off Baja California, the longest migration of any mammal. Scientists believe that gray whales may leave their homes to avoid predators such as killer whales, which feed on gray whale calves. In the

[11] Killer whales are generally considered to have only limited association with sea ice (Born 2000). However, they are regularly observed in polar seas, including in the Greenland-Iceland-Norwegian seas (Øien 1988) and near Barrow, Alaska, in summer (Moore and Huntington 2008). With sea ice reductions, killer whales may follow their fish and marine mammal prey that shift north into Arctic waters. With the rapid decline of sea ice in Hudson Strait, killer whales have begun to be seen with much greater frequency in far northern Canadian waters and are now regularly signed in Hudson Bay where they were historically absent (Higdon and Ferguson 2009).

North Pacific, they feed continuously on tiny crustaceans on the on the ocean floor, and off Baja, they breed, calve and socialize. Most gray whales do not eat during their swim south or when they arrive in Baja, so they must develop thick layers of blubber to help sustain them as they fast.

Whale watchers and marine scientists report that these whales have been delaying their southern journeys, pointing to climate change as the cause. Rising sea temperatures have disrupted the animals' home habitat in the waters between Alaska and Russia and the whales are spending more time in the north before they start their yearly trip south. In Monterey Bay, California where scientists have observed the whales for more than 20 years as they pass through, they have seen the whales passing through the bay a week to 10 days later. As warmer waters melt sea ice, other animals move into the whales' habitat and start feeding on the crustaceans or the crustacean's life cycle is altered or less productive due to climate change. The gray whales have to travel further north and feed longer to get their fill. These changes have disrupted the timing of the whales' yearly migration south. As the sea ice continues to thin, the gray whales' food supply will likely decrease.

Some gray whales are remaining in the Arctic year round (Moore and Huntington, 2008, Stafford et al., 2007). Gray whales are perhaps the most adaptable and versatile of the baleen whales as they are dynamic and opportunistic foragers (e.g., Nerini, 1984, Darling et al., 1998, Dunham and Duffus, 2001). Recently they have been documented feeding year-round off Kodiak, Alaska (Moore et al., 2007) and surprisingly researchers have detected their calls in the western Beaufort Sea throughout the winter of 2003–2004 (Moore et al., 2006). While there was a marked wintertime reduction in calling rates (Stafford et al., 2007), it is clear that a few gray whales remained in the Beaufort Sea over winter (Moore and Huntington, 2008). It is unknown whether or not the gray whales that remained in the Beaufort Sea found prey, but there could be a net metabolic advantage whereby the energetic costs of thermoregulation in cold water are offset by not undertaking the 10,000-km round-trip migration and remaining in northern seas to take advantage of spring forage (Moore and Huntington, 2008).

Evidence supports that the eastern North Pacific gray whale population has responded to climate change at ocean basin scales (Moore and Huntington, 2008). This population has demonstrated: (1) southbound migration timing shifting one week later, coincident with the late 1970s regime shift in the North Pacific Ocean (Rugh et al., 2001); (2) calving rates that are correlated positively with early-season ice-free conditions in the Chirikov Basin (Perryman et al., 2002); (3) lack of whales feeding in the Chirikov Basin coincident with a decline in benthic infauna (Moore et al., 2003); and (4) the previously mentioned detection of calls during winter northeast of Barrow, Alaska (Moore et al., 2006).

INCREASED HUMAN ACTIVITIES IN THE ARCTIC

Warmer temperatures and longer open-water seasons will facilitate increased human activities in the Arctic, including commercial shipping, commercial fishing, military activities, oil and gas operations, tourism, and coastal development (Ragen et al., 2008). These activities will pose multiple risks to whales through disturbance from human presence or noise, ship strikes, direct and indirect fishery interactions, exposure to contaminants, and loss or degradation of important habitat for reproduction and feeding. Such risks are expected to increase in severity over time as sea ice habitat declines and human activities encroach into previously inaccessible areas. Climate change induced impacts are likely to compound and exacerbate these threats by reducing resilience and adaptive capacity because of resource deployment to competing needs.

Trends in environmental conditions and human activities will vary regionally, as will their impacts on marine mammals (Simpkins, 2009). For example, climate change models predict that by 2050 summer sea ice will have declined significantly in the Chukchi, Beaufort, and Barents Seas, whereas it will remain relatively constant in Baffin Bay and the Labrador Sea (Overland and Wang, 2007). Changes in the type and intensity of human activities likely will reflect both changes in the accessibility of various regions and the abundance of their resources (Simpkins, 2009).

The Northwest Passage is the sea route that connects the Atlantic with the Pacific Ocean through Canada's high Arctic. It would save valuable time and fuel for ships that now travel through the Suez Canal in Egypt or the Panama Canal in Central America. In September 2007, the Northwest Passage was ice-free for the first time since satellite records began in 1978 and much earlier than predicted (BBC, 2007). At the same time, satellite images from the Northeast Passage along northern Siberia also showed that route to be almost clear of ice (BBC, 2007). Ships have transited both passages in the past—with difficulty—including in recent years as ice cover thinned. But this was the first time the Northwest Passage has been fully navigable since the European Space Agency began measuring the Arctic sea ice with satellites (BBC, 2007).

The Lancaster Sound region, in the eastern part of the Northwest Passage, is an important summer habitat and migratory corridor for belugas and narwhals (Elliott and Simmonds, 2007). This region, and the southern portion of the North Water polynya of northern Baffin Bay are among the richest areas for marine mammals and birds in the Canadian Arctic (Stirling, 1997). Reduced sea ice from climate change is making the Northwest Passage more easily navigable, allowing increased boat traffic for longer lengths of time (Elliott and Simmonds, 2007). This results in an increased risk of additional chemical and noise pollution, and more fatal collisions between whales and ships[12] (Elliott and Simmonds, 2007).

In addition, commercial fishing fleets may move north following changes in fish stock distribution, as well as retreating ice. Ice-free waters will make additional oil and gas exploration and development more financially possible. These industrial activities would further contribute to increased noise (from seismic exploration and drilling for example), increased vessel traffic and an increased risk of accidents such as oil spills.

Belugas have been reported to swim away quickly from icebreaking ships approaching within 35-50 km; narwhals have been reported to show less overt panic to approaching ships, although their underwater vocalizations stop temporarily (Finley et al., 1990a). Bowhead whales are perhaps the most sensitive of all large whales, reacting from seismic and shipping noise from over 30 km away (Richardson, 2006) and leaving feeding areas where there is industrial noise (Elliott and Simmonds, 2007). Like their relatives, the North Atlantic right whales (*Eubalaena glacialis*), it is reasonable to assume that bowheads will fare badly in the face of increased fishing activities (Burns, 1993).

While it is difficult to predict exactly what the direct and indirect biological and ecological consequences of these changes might be, experience from areas outside the Arctic (e.g., Northridge and Hofman, 1999, Laist et al., 2001, Reynolds et al., 2005, Bejder et al., 2006) suggests that these activities will pose new risks to marine mammals. In many cases, it will be impossible to fully mitigate the negative impacts of climate change; therefore increased efforts must be made to reduce all other human-induced threats.

---

[12] An analysis of 292 large whale - ship strikes around the world revealed that 198 (68%) were fatal and 48 (16%) resulted in injury to the animal (Jensen 2003). In most cases the fate of injured whales is not known.

SOUTHERN OCEAN – WHALES AND THEIR FOOD WEB

Over half of the world's marine mammal biomass, including around one fifth of the world's cetacean species are found in the Southern Ocean (Boyd, 2002). Most species of baleen whales and male sperm whales in the Southern Hemisphere migrate between low-latitude breeding grounds in winter and highly productive Antarctic feeding grounds in summer. Eight species of baleen whales feed almost entirely on Antarctic krill (*Euphausia superba*), tiny shrimp-like marine animals that depend on sea ice and are the region's abundant keystone prey species. These whales feed almost exclusively inside the rich Antarctic waters where they build their energy stores for their extensive migration and reproductive activities during the rest of the year (Simmonds & Elliott, 2009).

Krill's abundance is due to the delicate balance in the Southern Ocean between ice and oceanographic conditions, making food production possible at the lowest levels of the food chain. Krill assemblages occur along the ice edge (boundary between sea ice and open water) and transition zones (fronts between water masses of differing salinity and temperature) where their food source is the most plentiful.

The ice edge provides ideal habitat for krill and for the food upon which it relies; in summer, the ice edge is the site of massive algal blooms and in winter, krill larvae find a safe haven. In the fall, sea ice freezes along with the algae and microbes. These microscopic creatures live within the ice crystals throughout the long dark winter, feeding the maturing krill larvae. The summer sun melts the ice and releases the algae, microbes and young krill. Photosynthesis and continuous sunlight exposure provides the perfect recipe for a flourishing crop of the Southern Ocean food chain's lowest level. Krill and young fish multiply, making the ice edge the area of highest productivity in the Southern Ocean ecosystem. It provides a rich prey base and main foraging site for many whale species.

Transition zones, where different water masses meet, also provide essential habitat for the lower trophic levels. A water mass is a large water body with very similar characteristics. In different regions and depths, water differs in temperature, salinity and density. Water masses that contain older water, low oxygen levels and high nutrient amounts, move closer to the surface at transition zones. These conditions allow phytoplankton to grow which supports krill to grow and then whales and other species that rely on this important prey species. These fronts attract baleen whales which gorge on krill and sperm whales which feed on squid gather, indicating the ecosystem importance of these areas.

Many direct and indirect impacts to Southern Ocean whales are expected from climate change. One of the most significant impacts to resident and migratory whales from loss of sea ice is a reduction in their major food source, ice dependent Antarctic krill. Krill, small shrimplike zooplankton, are a key component of the Antarctic ecosystem because of their importance as a foundation prey species. Since the 1970s, krill populations have declined by as much as 80% in the Scotia Sea and northern Antarctic Peninsula due to loss of winter sea ice (Atkinson et al., 2004). The decline in krill has had serious implications for the Southern Ocean food webs in these areas, including penguins, albatrosses, seals and whales, all of which have wide foraging ranges but are susceptible to krill shortages (Atkinson et al., 2004). A 2001 study found that population size and reproductive performance were declining in land-based krill-eating predators such as seals and penguins in South Georgia, and concluded that "the biomass of krill… was sufficient to support predator demand in the 1980s, but not in the 1990s" (Reid and Croxall, 2001). In 2006 additional analyses revealed that southern right whales (*Eubalaena australis*) in South Georgia were experiencing similar impacts to the seals and penguins. The breeding success of the right whales was highly correlated with the impacts of global climate signals on krill, their main food source. The researchers concluded that "even quite small changes in oceanographic conditions in the Southern Ocean could affect southern right whale population dynamics" (Leaper et al., 2006).

Along the West Antarctic Peninsula, the distribution of humpback and minke whales is closely linked to the boundary between frozen sea ice and open water – the "ice edge" (Thiele et al., 2004; Friedlaender et al., 2006). Since 1980, the Antarctic Peninsula region has experienced the greatest warming of the Southern Hemisphere (Overland et al., 2008). Climate models project that this region will continue to rapidly lose sea ice, and consequently, lose important foraging habitat for humpback and minke whales (Tynan and Russell, 2008). In the Atlantic Sector, Scotia Arc, the loss of sea ice coverage could affect the movements and foraging of minke whales and blue whales, which appear to migrate through open waters to the ice-edge zone to feed (Reilly et al., 2004). Averaging across all models, it is projected that whales in the Atlantic Sector, approaching the Weddell Sea, would encounter a 10-20% reduction in ice coverage at the time of 2°C warming, although one model projected a reduction in ice cover of up to 40% (Tynan and Russell, 2008).

Critical whale habitats are also projected to move south to colder waters (Tynan and Russell, 2008). This includes productive frontal zones - boundaries between different water masses where nutrient-rich upwelling stimulates phytoplankton growth – that support substantial prey species populations for whales (Tynan and Russell, 2008). Under 2°C global warming, Southern Ocean fronts are projected to move southward by 2-5° latitude (about 200-500 km) (Tynan and Russell, 2008). Humpback and blue whales will have to travel extra 200-500 km farther south on their migration to feed on krill in these rich waters to build up reserves to for the rest of the year (Tynan and Russell, 2008). These species are only now coming back from the brink of extinction after populations were decimated by commercial whaling, mainly during the first half of the 20th century. Traveling further for food will likely involve a greater expenditure of energy and allow for less feeding time upon arrival (Tynan and Russell, 2008). Moreover, the overall foraging habitat available will be reduced because the frontal zones will move closer together as they move south, compressing the space between them and reducing the area of valuable foraging habitat for migratory whales, including the blue whale, humpback whale, fin whale and sperm whale (Tynan and Russell, 2008; Tynan, 1998; Tynan, 1997). As the krill is dependent on sea ice, less sea ice is also expected to reduce the abundance of food for whales in the feeding areas (Tynan and Russell, 2008).

Changes in Antarctic sea ice play a major role in climate processes with the annual formation and loss of approximately 15 million km^2 of sea ice affecting global ocean circulation, particularly by forming dense bottom water (Nicol et al., 2008). Climate change will likely impact the sea ice cycle by changing the Antarctic region's productivity (Nicol et al., 2008). The sea ice is critical habitat, supporting many species of wildlife, particularly large vertebrates such as seals, seabirds and whales, some which have been exploited to near extinction (Nicol et al., 2008). Southern Ocean cetacean species will be directly impacted by changes in sea ice patterns as well as indirectly by changes in their principal prey, Antarctic krill (Nicol et al., 2008). Basically, less sea ice means fewer krill and fewer krill could mean fewer baleen whales. Consequently, scientists have called for the initiation of long-term studies to examine variability and change at all trophic levels in the Southern Ocean ecosystem to reduce uncertainty and to improve predictive power (Nicol et al., 2008).

Recommended Actions

Climate change will have adverse and possibly irreversible repercussions on high latitude sea ice ecosystems. Loss of sea ice habitat and the productive food web associated with it and increasing human presence and activities are likely the most significant threats to ice-dependent marine mammals, including cetaceans. Cumulative effects of these and other factors may well include extirpation of populations and even extinction of marine mammal species. Measures to minimize or mitigate the effects of climate change on ice-dependent marine mammals are few and, at best, can address only the secondary effects of climate change, including those resulting from additional human presence and activities. As part of a larger strategy, such measures may make the difference between persistence and extirpation or, in the worst case, extinction. Prevention through reductions in greenhouse gas emissions appears to be the only approach that can ensure the long-term conservation of Arctic marine mammals and Arctic ecosystems as we know them (Ragen et al., 2008).

We face a conundrum regarding climate change impacts to cetaceans. On one hand, we know climate change is occurring and will continue even if governments act conservatively now to restrict greenhouse gas emissions. In many cases, there are significant information gaps about what we know about cetacean biology, habitat and prey needs. Moreover, we do not know with certainty what all of the climate change impacts will be on cetaceans, their habitat and prey. We do not know what measures society can take to modify or stop the physical manifestations of climate change such as melting sea ice and permafrost, increasing temperature and rainfall, altered distribution of species and their prey (Ragen et al., 2008).

Need for Monitoring and Modeling

Given the uncertainty, it is necessary to collect baseline data about different cetacean species and their prey and continue to monitor changes. Monitoring marine mammal populations and stocks will be essential to provide an empirical measure of their response to the cumulative impact of all risk factors, natural and human-caused. To date, studies of high latitude marine mammal stocks generally have been insufficient to determine stock structure, status, trends, or the influence of individual risk factors on status and trends

(Ragen et al., 2008). The current funding support for research limits informed decision making about the status of marine mammals now and, if not remedied, will undermine informed decision making in the future (Ragen al., 2008). Further, management efforts must increase the resilience of ecosystems and species to climate change as well as address changes in human activities secondary to climate change (e.g., shipping, tourism, oil and gas activities, fishing, development) (Ragen et al., 2008). Such measures may be essential in the short term to prevent losses from factors other than climate change and also as interim measures to retain species or distributions until longer-term measures take effect (Ragen et al., 2008).

In view of the time lag between climate change cause and consequence, the uncertainty regarding full consequences and the potential severity of those consequences, climate change must be addressed with proactive, adaptive, and precautionary management aimed at prevention of adverse effects (Ragen et al., 2008). It is a matter of urgency that long-term studies are initiated to examine variability and change at all trophic levels in the Arctic and Southern Ocean ecosystems to reduce uncertainty and to improve predictive power (Nicol et al., 2008). Modeling studies need to take into account past and future harvesting, inter-annual variability as well as the longer term trends that will occur as a result of global warming, but they will have to be built on a far more robust understanding of the structure and functioning of the marine ecosystem (Nicol et al., 2008).

EMISSIONS REDUCTIONS

It is clear that the world must urgently make dramatic changes in order to avoid irrevocable consequences of dangerous climate change by limiting global mean temperature rise to well below 2°C above pre-industrial levels (Tynan and Russell 2008). Reducing energy demand, improving energy efficiency, deploying renewable energy and other low-carbon technologies, stopping and reversing loss and degradation of forests and prairies are all crucial elements to keeping global warming to below 2°C. Action is needed by governments of the world to agree to targets, to collaborate on effective strategies, and to influence and coordinate the investment of the massive funds necessary so that future needs are met safely and sustainably.

Adaptation

While making efforts to slow down climate change, it is critically important that climate change considerations be incorporated into conservation plans, assessments and strategies for cetaceans to improve the resilience of ecosystems and species to climate change (Simmonds and Isaac 2007). This could be achieved through three main principles (Hansen et al., 2003):

1. Protection of adequate and appropriate space. This should include the protection of habitats critical for breeding and feeding, and the protection of climate refugia – those areas that are less vulnerable to changes in climate than others. In designing protected areas, proactive planning must done to determine how climate induced factors may change the geography of the most important attributes to be protected. For example, if cetaceans change their distributions and establish new critical habitat areas, conservation and management efforts will have to move with them.
2. Limit all non-climate stresses. There are many on whales and the marine environment, and climate change will have a synergistic effect on these. Non-climate stressors can be locally controlled (i.e., pollution, fishing, noise); as such, increased efforts must be made to reduce all these threats.
3. Adaptive management. Given the uncertainty about the exact nature of impacts of climate change on whales and their responses to it, a responsive and flexible approach is required, combined with rigorous monitoring (Simmonds and Elliott 2009).

Developing a Sensitivity / Vulnerability Index for Cetaceans

Sensitivity or vulnerability indices can be useful tools for designing and assessing adaptation actions, as well as for identifying key research needs (Hoffman et al. 2009). Building on the work of Laidre et al (2008), who developed sensitivity indicators for Arctic marine mammals, Mark Simmonds has proposed a draft set of sensitivity indicators for cetaceans as a whole. Indicators included population characteristics such as genetic variability, geographic distribution, and reproductive rates as well as physical and ecosystem level changes (Hoffman et al., 2009).

References

ACIA. 2004. *Impacts of a warming Arctic: Arctic Climate Impact Assessment.* Cambridge University Press.

ASOC. XXXII Antarctic Treaty Consultative Meeting, April 6-17, 2009. Policy Implications Arising from SCAR's Report: Antarctic Climate Change and the Environment *(http://www.asoc.org/Portals/0/policy%20implications%20SCAR%20report.pdf)*.

Ainley, D.G., Clarke, E.D., Arrigo, K., Fraser, W.R., Kato, A., Barton, K.J. and P.R. Wilson. 2005. Decadal-scale changes in the climate and biota of the Pacific sector of the Southern Ocean, 1950s to the 1990s. *Antarctic Science*, 17,171-182.

Alexander, V. 1981. Ice-biota interactions: An overview. In: Hood, D.W., and Calder, J.A., eds. The eastern Bering Sea Shelf: *Oceanography and resources.* Vol. 2. Seattle: University of Washington Press. 757– 761.

Arrigo, K. R., and G. L. van Dijken. 2004. Annual cycles of sea ice and phytoplankton in Cape Bathurst polynya southeastern Beaufort Sea, Canadian Arctic. *Geophysical Research Letters* 31:L08304.

Atkinson, A., Siegel, V., Pakhomov, E. And Rothery, P. 2004. Longterm decline in krill stock and increase in salps within the Southern Ocean. *Nature*. 432, 100-103.

BBC. "Warming 'opens Northwest Passage'" September 14, 2007. *(http://news.bbc.co.uk/2/hi/americas/6995999.stm).*

Bannister, J.L. 2002. Baleen whales. *In: Encyclopedia of marine Mammals.* (eds W.F. Perrin, B. Würsig and J.G.M. Thewisse). P 62-72, San Diego: Academic Press.

Bass C.L., Simmonds M.P., Isaac S.J. 2006. *An overview of the potential consequences for cetaceans of oceanic acidification.* Paper presented to the Scientific Committee of the International Whaling Commission, SC/58/E10.

Bejder, L., Samuels, A., Whitehead, H., Gales, N., Mann, J., Connor, R., Heithaus, M., Watson-Capps, J., Flaherty, C. and Krutzen, M. 2006. Decline in relative abundance of bottlenose dolphins exposed to long-term disturbance. Conservation Biology 20:1791–1798.

Bluhm, B.A., and R. Gradinger. 2008. Regional variability in food availability for Arctic marine mammals. *Ecological Applications* 18(2) Supplement: S77-S96.

Born, E. W. 2000. Marine mammals. Pages 185–202 *in* Born, E. W., and J. Bocher. The ecology of Greenland. *Schmidt Grafisk,* Vojens, Nuuk, Greenland.

Boyd, I.L. 2002. Antarctic Marine Mammals. *Encyclopedia of Marine Mammals.* Academic Press, San Diego.

Bradstreet, M.S.W., Finley, K.J., Sekerak, A.D., Grifiths, W.B., Evans, C.R., Fabijhan, M.F., and Stallard, H.E. 1986. Aspects of the biology of arctic cod (*Boreogadus saida*) and its importance in arctic marine food chains. *Canadian Technical Report of Fisheries and Aquatic Sciences* No. 1491. 193 p.

Branch, T. A., Stafford, K. M., Palacios, D. M., Allison, C., Bannister, J. L., *et al.* 2007. Past and present distribution, densities and movements of bluewhales *Balaenoptera musculus* in the Southern hemisphere and northern Indian Ocean. *Mammal Review* 37(2), 116–175.

Brierley, A.S. and Thomas, D.N. 2002. On the ecology of the Southern Ocean pack ice. Advances in Marine Biology 43: 171-278.

Burns, J.J., Montague, J.J., Cowles, C.J. 1993. *The bowhead whale.* Allen Press, Lawrence, Kansas. 787 pp.

Cabal, J. A., F. Alvarez-Marque´s, J. L. Acun˜a, M. Quevedo, R. Gonzalez-Quiro´s, I. Huskin, D. Ferna´ndez, C. R. Del Valle, and R. Anado´n. 2002. Mesozooplankton distribution and grazing during the productive season in the Northwest Antarctic Peninsula. *Deep-sea Res. II* 49: 869–882.

Canadell JG, *et al.* 2007. Contributions to accelerating atmospheric CO2 growth from economic activity, carbon intensity, and efficiency of natural sinks. *Proc Nat Acad Sci USA* 104:18866–18870.

Crawford, R.E., and Jorgensonn, J.K. 1993. Schooling behaviour of arctic cod, *Boreogadus saida*, in relation to drifting pack ice. *Environmental Biology of Fishes* 36:345–357.

Daly, K.L. 1990. Overwintering development, growth and feeding of larval *Euphausia superba* in the Antarctic marginal ice zone. *Limnology and Oceanography* 35: 1564-1576.

Daly, K.L. and Macaulay, M.C. 1991. Influence of physical and biological mesoscale dynamics on the seasonal distribution and behavior of *Euphausia superba* in the Antarctic marginal ice zone. *Mar. Ecol. Prog. Ser.* 79: 37-66.

Darling, J. D., K. E. Keogh, and T. E. Steeves. 1998. Gray whale (*Eschrichtius robustus*) habitat utilization and prey species off Vancouver Island, B.C. *Marine Mammal Science* 14:692–720

Davis, R.A., Finley, K.J., and Richardson, W.J. 1980. The present status and future management of arctic marine mammals in Canada. Report by LGL Ltd. for the Science Advisory Board of the Northwest Territories, *Yellowknife*. Report 3. 93 p.

Dayton, P. K., S. Thrush, and F. C. Coleman. 2002. Ecological effects of fishing in marine ecosystems in the United States. *Pew Commission Report. Pew Oceans Commission,* Arlington, Virginia, USA.

Doney, S.C. 2006. The dangers of ocean acidification. *Scientific* American 294, 58-65.

Ducklow, H.W., Baker, K., Martinson, D.G., Quetin, L.B., Ross, R.M., Smith, R.C., Stammerjohn, S.E., Vernet, M. and W. Fraser. 2007. Marine pelagic ecosystems: the West Antarctic Peninsula. *Philosophical Transactions of the Royal Society* B, 362, 67–94.

Dunbar, M.J. 1981. Physical causes and biological significance of polynyas and other open water in sea ice. In: Stirling, I., and Cleator, H., eds. Polynyas in the Canadian Arctic. *Canadian Wildlife Service Occasional Paper Number 45*. 29–43.

Dunham, J. S., and D. A. Duffus. 2001. Foraging patterns of gray whales in central Clayquot Sound BC, Canada. *Marine Ecology Progress Series* 223:299–310

Edwards, M., Johns, D.G., Licandro, P., John. A.W.G. and Stevens, D.P. 2007. *Ecological Status Report:* results from the CPR survey 2005/2006. SAHFOS Technical Report 4, 1–8. Plymouth, UK, ISSN 1744-0750.

Elliott, W. and Simmonds, M. 2007. Whales in Hot Water? The Impact of a Changing Climate on Whales, Dolphins and Porpoises: *A call for action.* WWF-International, Gland Switzerland / WDCS, Chippenham, UK.

Feely, R.A. et. al., Impact of Anthropogenic CO_2 on the $CaCO_3$ System in the Oceans 2004. Science 305 (5682), 362.

Field, C.B., Behrenfeld, M.J., Randerson, M.T. and Falkowski, P. 1998. Primary Production of the Biosphere: Integrating Terrestrial and Oceanic Components Science 281 (5374), 237.

Finley, K.J., Bradstreet, M.S.W., and Miller, G.W. 1990. Summer feeding ecology of harp seals (*Phoca groenlandica*) in relation to arctic cod (*Boreogadus saida*) in the Canadian High Arctic. *Polar Biology* 10:609–618.

Finley, K.J., Miller, G.W., Davis, R.A., and Greene, C.R. 1990a. Reactions of Belugas, Delphinapterus leucas, and narwhals, Monodon monoceros, to ice-breaking ships in the Canadian High Arctic. *Canadian Bulletin of Fisheries and Aquatic Sciences*. 224, 97-117.

Friedlaender, A.S., Halpin, P.N., Qian, S.S., Lawson, G.L., Wiebe, P.H., Thiele, D. and Read, A.J. 2006. Whale distribution in relation to prey abundance and oceanographic processes in shelf waters of the Western Antarctic Peninsula. Mar. Ecol. *Prog. Ser.* 317: 297-310.

George, J. C., J. Zeh, R. Suydam, and C. Clark. 2004. Abundance and population trend (1978–2001) of western Arctic bowhead whales surveyed near Barrow, Alaska. *Marine Mammal Science* 20:755–773.

Gloersen, P., Campbell, W. J., Cavalieri, D. J., Comiso, J. C., Parkinson,C. L., and Zwally, H. J. 1992. Arctic and Antarctic Sea Ice 1978–1987: *Satellite Passive-microwave Observations andAnalysis*.(National Aeronautics and Space Administration: Washington, D.C.).

Hansen, L., Biringer, J.L. and Hoffman, J.R. 2003. Buying Time: A User's Manual for Building Resistance and Resilience to Climate Change in Natural Systems. *WWF*.

Heide-Jørgensen, M. P., Richard, P., Ramsay, M. and Akeeagok, S. 2002. Three recent ice entrapments of Arctic cetaceans in West Greenland and the eastern Canadian High Arctic. *NAMMCO Scientific Publications* 4:143–148.

Higdon, J.W. and S.H. Ferguson. 2009. Loss of Arctic sea ice causing punctuated change in sightings of killer whales (Orcinus orca) over the past century. *Ecological Applications* 19(5): 1365-1375.

Hoffman, J.R., Fonseca, A., and Drews, C. (eds). 2009. Cetaceans and Other Marine Biodiversity of the Eastern Tropical Pacific: Options for Adapting to Climate Change. Report from a workshop held February 9-11, 2009. *MINAET/WWF/EcoAdapt/CI/IFAW/TNC/WDCS/IAI/ PROMAR,* San Jose, Costa Rica.

Holland, M.M., Cecilia, M.B., Tremblay, B. 2006. Future abrupt reductions in the summer Arctic sea ice. *Geophysical research letters*. 33, L23503.

Huntington, H.P., Moore, S.E. 2008. Assessing the impacts of climate change on Arctic Marine Mammals. Ecological Applications: Vol. 18, Supplement: *Arctic Marine Mammals*, pp. S1-S2.

Hunt, G. L., P. Stabeno, G. Walters, E. Sinclair, R. D. Brodeur, J. M. Napp, and N. A. Bond. 2002. Climate change and control of the southeastern Bering sea pelagic ecosystem. *Deep-Sea Research* II 49:5821–5853.

IPCC 2007. Climate Change 2007: The Physical Science Basis. Summary for Policy makers. *Intergovernmental Panel on Climate Change.*

IPCC. 2007b. Climate Change 2007: Impacts, Adaptation and Vulnerability. Summary for Policymakers. *Intergovernmental Panel on Climate Change.*

IUCN Cetacean update of the 2008 IUCN Red List of Threatened Species, *(http://cmsdata.iucn.org/downloads/cetacean_table_for_website.pdf)*

IWC. 1997. Report of the IWC workshop on climate change and cetaceans. *Report of the International Whaling Commission* 47, 293-331.

Ishimatsu, A., Kikkawa, T., Hayashi M., Lee, .K.S. & Kita, J. 2004. Effects of CO2 on marine fish: larvae and adults. *Journal of Oceanography* 60, 731–741

Jacobs, S. 2006. Observations of change in the Southern Ocean. *Philosophical Transactions of the Royal Society A,* 364, 1657-81.

Jensen, A.S. and G.K. Silber. 2003. *Large whale ship strike database.* U.S. Department of Commerce, NOAA Technical Memorandum. NMFS-OPR-, 37 pp.

Kasamatsu, F., Ensor, P., and Joyce, G. G. 1998. Clustering and aggregations of minke whales in Antarctic feeding grounds. *Marine Ecology Progress Series* 168, 1–11.

Kasamatsu, F., Matsuoka, K., and Hakadama, T. 2000. Interspecific relationships in density among the whale community in the Antarctic. *Polar Biology* 23, 466–473.

Kawamura, A. 1994. A review of baleen whale feeding in the Southern Ocean. *Report of the International Whaling Commission* 44, 261–271.

Laidre, K. and Heide-Jørgensen, M.P. (retrieved November 1, 2009) "The Biology and Ecology of Narwhals". National Oceanic and Atmospheric Administration. *http://oceanexplorer.noaa.gov/explorations/06arctic/background/biology/biology.html.*

Laidre, K.L & Heide-Jørgensen, M.P. 2005. Arctic sea ice trends and narwhal vulnerability. *Biological Conservation,* 121, 509-517.

Laidre, K.L., Stirling, I., Lowry, L.F., Wiig, Ø., Heide-Jørgensen, M.P., and Ferguson, S.H. 2008. Quantifying the sensitivity of Arctic marine mammals to climate induced habitat change. *Ecological Applications 18*(Supplement):S97–S125.

Laist, D.W., Knowlton, A.R. Mead, J.G. Collet, A.S. and Podesta, M. 2001. Collisions between ships and whales. *Marine Mammal Science* 17:35–75.

Lalli, C. M., and R. W. Gilmer. 1989. *Pelagic Snails: The Biology of Holoplanktonic Gastropod Mollusks,* Stanford University Press, Stanford, CA, USA.

Le Quéré, C. Rödenbeck, C. Buitenhuis, E.T. Conway, T.J. Langensfelds, R. Gomez, A. Labuschangne, C. Ramonet, M. Nakazawa, T. Metzl, N. Gillett, N.P. Heimann, M. 2007. Saturations of the Southern Ocean CO_2 sink due to recent climate change. *Science* 316, 1735.

Leaper, R. Cooke, J., Trathan, P., Reid, K., Rowntree, V. and Payne R. 2006. Global climate drives southern right whale (*Eubalaena australis*) population dynamics. *Biol. Lett.*

Learmonth, J.A., Maclead, C.D., Santos, M.B., Pierce, G.J., Crick, H.Q.P., Robinson, R.A. 2006. Potential effects of climate change on marine mammals. *Oceanography and Marine Biology: An Annual Review.* 44, 431-464.

Lonne, O.J., and Gulliksen, B. 1989. Size, age and diet of polar cod, *Boreogadus saida* (Lepechin 1773), in ice covered waters. *Polar Biology* 9:187–191.

Lowry, L.F. 1993. Foods and feeding ecology. In: Burns, J.J., Montague, J.J., and Cowles, C.J., eds. The bowhead whale. Special Publication Number 2. *Lawrence, Kansas: The Society for Marine Mammalogy*. 201–238.

Lowry, L. 2000. Marine mammal–sea ice relationships. Pages 91–96 in H. P. Huntington, editor. Impacts of changes in sea ice and other environmental parameters in the Arctic. Report of the Marine Mammal Commission Workshop, 15–17 February 2000, Girdwood, *Alaska. Marine Mammal Commission*, Besthesda, Maryland, USA.

Lydersen, C., Angantyr, L.A., Wiig, Ø., and Øritsland, T. 1991. Feeding habits of Northeast Atlantic harp seals (*Phoca groenlandica*) along the summer ice edge of the Barents Sea. *Canadian Journal of Fisheries and Aquatic Sciences* 48: 2180–2183.

MacLeod, C., Bannon, S.M., Pierce, G.J., Schweder, C., Learmonth, J.A., Herman, J.S., Reid, R.J. 2005. Climate change and the cetacean community of north-west Scotland. *Biological Conservation.* 124, 477-483.

McCarthy, J. J., O. F. Canziani, N. A. Leary, D. J. Dokken, and K. S. White, editors. 2001. Climate change 2001: impacts, adaptability, and vulnerability. *Contribution of Working Group II to the Third Assessment Report of the Intergovernmental Panel on Climate Change.* United Nations Environmental Program. Cambridge University Press, Cambridge, UK.

Macdonald, R. W., T. Harner, and J. Fyfe. 2005. Recent climate change in the Arctic and its impact on contaminant pathways and interpretation of temporal data. *Science of the Total Environment* 342:5–86.

Moore, S.E., and Reeves, R.R. 1993. Distribution and movement. In: Burns, J.J., Montague, J.J., and Cowles, C.J., eds. The bowhead whale. Special Publication Number 2. Lawrence, Kansas: Society for Marine *Mammalogy*. 313– 386.

Moore, S.E. 2000. Variability of cetacean distribution and habitat selection in the Alaskan Arctic, Autumn 1982-91. *Arctic. Vol.* 53. No. 4, p. 448-460.

Moore, S.E., DeMaster, D.P., Dayton, P.K. 2000. Cetacean Habitat Selection in the Alaskan Arctic during Summer and Autumn. *Arctic*. Vol 53, No. 4, p 432-447.

Moore, S. E., J. M. Waite, L. L. Mazzuca, and R. C. Hobbs. 2000. Mysticete whale abundance and observations on prey association on the central Bering sea shelf. *Journal of Cetacean Research and Management* 2:227–234.

Moore, S. E., J. M. Waite, N. A. Friday, and T. Honkalehto. 2002. Cetacean distribution and relative abundance on the central-eastern and the southeastern Bering Sea shelf with reference to oceanographic domains. *Progress in Oceanography* 55:249–261.

Moore, S. E., J. M. Grebmeier, and J. R. Davies. 2003. Gray whale distribution relative to forage habitat in the northern Bering Sea: current conditions and retrospective summary. *Canadian Journal of Zoology* 81:734–742.

Moore, S.E., Laidre, K.L. and Overland, J.E. 2006. Trends in sea ice cover within habitats used by bowhead whales in the Western Arctic. *Ecological Applications:* 16, (3) 932-944.

Moore, S. E., K. M. Wynne, J. Clement, and J. M. Grebmeier. 2007. Gray whale occurrence and forage southeast of Kodiak Island, Alaska. *Marine Mammal Science* 23:419–428.

Moore, S.E. and Huntington, H.P. 2008. Arctic marine mammals and climate change: impacts and resilience. *Ecological Applications* 18(Supplement):S157–S165.

Murase, H., Matsuoka, K., Ichii, T., and Nishiwaki, S. (2002). Relationship between the distribution of euphausiids and baleen whales in the Antarctic (35"E–145"W). *Polar Biology* 25, 135–145.

National Snow and Ice Data Center, United States. *http://nsidc.org/*

Nerini, M. 1984. A review of gray whale feeding ecology. Pages 423–450 in Jones, M. L., S. L. Swartz, and S. Leatherwood. The gray whale, *Eschrichtius robustus*. Academic Press, New York, New York, USA.

Nicol, S., Worby, A. and Leaper, R. 2008. Changes in the Antarctic sea ice ecosystem: potential effects on krill and baleen whales. *Marine and Freshwater Research* 59, 361–382.

Northridge, S. P., and Hofman, R.J. 1999. *Marine mammal interactions with fisheries.* Pages 120–155 *in* Twiss, J. R.Jr., and R. R. Reeves. Conservation and management of marine mammals. Smithsonian Institution Press, Washington, D.C., USA.

Øien, N. 1988. The distribution of killer whales (*Orcinus orca*) in the North Atlantic based on Norwegian catches, 1938-1981, and incidental sightings, 1967-1987. Rit Fisk. 11: 65-78.

Øien, N. 2003. Report of the Norwegian 2003 survey for minke whales in the Svalbard area. Scientific Committee Report SC/55/O8. *International Whaling Commission,* Cambridge, UK.

Orr J.C., Fabry V.J., Aumont O., Bopp L., Doney S.C., Feely R.A., Gnanadesikan A., Gruber N., Ishida A., Joos F., Key R.M.,Lindsay K., Maier-Reimer E., Matear R., Monfray P., Mouchet A., Najjar R.G., Plattner G.K., Rodgers K.B., Sabine C.L., Sarmiento J.L., Schlitzer R., Slater R.D., Totterdell I.J., Weirig M.F., Yamanaka Y. and Yool A. 2005. Anthropogenic ocean acidification over the twenty-first century and its impact on calcifying organisms. *Nature* 437, 681–686.

Overland, J. E., and P. J. Stabeno. 2004. Is the Climate of Bering Sea Warming and Affecting the Ecosystem?, Eos Trans. AGU, 85(33).

Overland, J.E., and M. Wang, 2007: Future regional Arctic Sea ice declines. *Geophys. Res. Lett.*, 34, L17705.

Overland, J., Turner, J., Francis, J., Gillett, N., Marshall, G., and Tjernström, M. 2008. The Arctic and Antarctic: *Two faces of climate change.* EOS 89(19): 177-178.

Parkinson, C. L., D. J. Cavalieri, P. Gloersen, H. J. Zwally, and J. C. Comiso. 1999. Arctic ice extents, areas and trends, 1978–1996. *Journal of Geophysical Research* 104:20837–20856.

Perryman, W. L., M. A. Donahue, P. C. Perkins, and S. B. Reilly. 2002. Gray whale calf production 1994–2000: Are observed fluctuations related to changes in seasonal ice cover. Marine Mammal Science 18:121–144.

Prentice et al., in Climate Change 2001: *The Scientific Basis, Contribution of Working Group I to the Third Assessment Report of the Intergovernmental Panel on Climate Change,* J. Houghton et al., Eds. (Cambridge Univ. Press, New York, 2001), pp. 183–238.

Ragen, T.J., Huntington, H.P., Hovelsrud, G.K. 2008 Conservation of Arctic Marine Mammals Faced With Climate Change. Ecological Applications: Vol. 18, Supplement: *Arctic Marine Mammals*, pp. S166-S174.

Raven et al., 2005 *Ocean acidification due to increasing atmospheric carbon dioxide.* The Royal Society Policy Document 12/05, London.

Reid, K. and Croxall, J.P. 2001. *Environmental response of upper trophic-level predators reveals a system change in an Antarctic marine ecosystem.* Proc. R. Soc. Lond. 268, 377-384.

Reilly, S., Hedley, S., Borberg, J., Hewett, R., Thiele, D., Watkins, J. and Naganobu, M. 2004. Biomass and energy transfer to baleen whales in the South Atlantic Sector of the Southern Ocean. *Deep-Sea Res.* II 51: 1397-1409.

Reynolds, J.E.III, Perrin, W.F. Reeves, R.R., Montgomery, S. and Ragen, T.J. 2005. *Marine mammal research: conservation beyond crisis.* The Johns Hopkins University Press, Baltimore, Maryland, USA.

Richardson W.J. (ed). 2006. *Monitoring of industrial sounds, seals, and bowheads whales, near BP's Northstar Oil Development, Alaskan Beaufort Sea,* 1999-2004. [Updated Comprehensive Report, April 2006.]

Riebesell, U. et al., 2000. Reduced calcification of marine plankton in response to increased atmospheric CO_2. *Nature* 407, 364-367.

Rignot, E., and P. Kanagaratnam. 2006. Changes in the velocity structure of the Greenland ice sheet. *Science* 311:986–990.

Rosa, R. and Seiebel, B.A. 2008. Synergistic effects of climate-related variables suggest future physiological impairment in a top oceanic predator *Proceedings of the National Academy of Sciences.*

Rugh, D. J., K. E. W. Shelden, and A. Schulman-Janiger. 2001. Timing of the gray whale southbound migration. *Journal of Cetacean Research and Management* 3:31–39.

Schmittner, A. 2005. Decline of the marine ecosystem caused by a reduction in the Atlantic overturning circulation. *Nature* 343:628–633.

Shepherd, A., D. Wingham, T. Payne, and P. Skvarca. 2003. Larsen ice shelf has progressively thinned. *Science* 302:856–859.

Seibel B.A. and Dierssen H.M. 2003. Tip of the iceberg: Cascading trophic impacts of reduced biomass in the Ross Sea, *Antarctica. Biological Bulletin* 205, 93–97.

Seibel, B.A., and Fabry, V. J.,in *Climate Change and Biodiversity: Synergistic Impacts,* L. Hannah, T. Lovejoy, Eds. (Conservation International, Washington, DC, 2003), pp. 59–67.

Siegstad, H., and M. P. Heide-Jørgensen. 1994. Ice entrapments of narwhals (Monodon monoceros) and white whales (Delphinapterus leucas) in Greenland. *Meddeleser om Grønland Bioscience* 39:151–160.

Simmonds, M. and Elliott, W. 2009. Climate change and cetaceans: concerns and recent developments. *Journal of the Marine Biological Association of the United Kingdom,* 89(1), 203–210.

Simmonds, M. and Isaac, S. 2007. The impacts of climate change on marine mammals: early signs of significant problems. *Oryx.* 41, (0), 1-8.

Simpkins, M. 2009. *A framework for monitoring Arctic marine mammals.* CAFF CBMP Report.

Stafford, K. M., S. E. Moore, and M. Spillane. 2007. Gray whale calls recorded near Barrow, Alaska, throughout the winter of 2003–04. *Arctic* 60:167–172.

Stirling, I. 1997. The importance of polynyas, ice edges, and leads to marine mammals and birds. *Journal of Marine Systems.* 10, 9-21.

Stirling, I., Lunn, N.J., Iacozza, J., Elliott, C. and Obbard. M. 2004. Polar bear distribution and abundance on the southwestern Hudson Bay coast during the open water season, in relation to population trends and annual ice patterns. *Arctic* 57:15–26.

Stirling, I., and T. G. Smith. 2004. Implications of warm temperatures and an unusual rain event for the survival of ringed seals on the coast of southeastern Baffin Island. *Arctic* 57:59–67.

TRS (The Royal Society). 2005. *Ocean Acidification Due to Increasing Atmospheric Carbon Dioxide.* The Royal Society, London, UK.

Thiele, D.C., Chester, E.T., Moore, S.E., Sirovic, A., Hildebrand, J.A., and Friedlaender, A.S. 2004. Seasonal variability in whale encounters in the West Antarctic Peninsula. *Deep-Sea Res.* II 51: 2311-2325.

Thomas, C. D., et al. 2004. Extinction risk from climate change. *Nature* 427:145–148.

Turley, C., Blackford, J., Widdicombe, S., Lowe, D., Nightingale, P.D. & Rees, A.P. 2006. Reviewing the impact of increased atmospheric CO2 on oceanic pH and the marine ecosystem. *In: Avoiding Dangerous Climate Change*, Schellnhuber, H J., Cramer,W., Nakicenovic, N., Wigley, T. and Yohe, G (Eds), Cambridge University Press, 8, 65-70.

Turner, J., Lachlan-Cope, T.A., Colwell, S., Marshall, G.J., Connolley, M. 2006. *Significant Warming of the Antarctic Winter Troposphere.*

Tynan, C.T. 1997. Cetacean distributions and oceanographic features near the Kerguelen Plateau. *Geophys. Res. Lett.* 24: 2793-2796.

Tynan, C.T. 1998. Ecological importance of the southern boundary of the Antarctic Circumpolar Current. *Nature* 392: 708-710.

Tynan, C. T. 2004. Cetacean populations on the SE Bering Sea shelf during the late 1990s: implications for decadal changes in ecosystem structure and carbon flow. *Marine Ecology Progress Series* 272:281–3000.

Tynan, C.T. and DeMaster, D.P. 1997. Observations and Predictions of Arctic Climate Change: Potential Effects on Marine Mammals. *Arctic*. 50, (4) 308-322.

Tynan, C.T. and Russell, J.L. 2008. *Assessing the impacts of future 2C global warming on Southern Ocean cetaceans*. International Whaling Commission, Scientific Committee document SC/60/E3.

Walsh, J. E. 2008. Climate of the Arctic marine environment. *Ecological Applications* 18(Supplement):S3–S22.

Welsh, H.E., Bergmann, M.A., Siferd, T.D., Martin, K.A., Curtis, M.F., Crawford, R.E., Conover, R.J., and Hop, H. 1992. Energy flow through the marine ecosystem of the Lancaster Sound Region, Arctic Canada. *Arctic* 45: 343 –357.

Cetaceans and Other Marine Biodiversity of the Eastern Tropical Pacific: Options for Adapting to Climate Change, 2009. IWC/61/18 (*http://www.iwcoffice.org/_documents/commission/IWC61docs/iwc61docs .htm*).

In: Whales and Dolphins
Editor: Craig A. Murray, pp. 89-112 ISBN 978-1-61668-693-2

Chapter 3

ECOLOGICAL AND EVOLUTIONARY DIVERGENCE IN WHALES AND DOLPHINS

Andrew D. Foote[1*,2•], M. Thomas, P. Gilbert[2] and Jennifer A. Jackson[3♦]

1. Institute of Biological and Environmental Sciences, University of Aberdeen, School of Biological Sciences, UK
2. Natural History Museum of Denmark, University of Copenhagen, Denmark
3. Marine Research Limited, Portsmouth, Hampshire, UK

Speciation is the process by which populations of an ancestral species are subject to divergent selection that leads to genetic and phenotypic differences and reproductive isolation between populations and ultimately the formation of new species. Ecology has long been considered as having an integral role in this process as the major force behind divergent selection (Mayr 1942; Dobzhansky 1951; Endler 1977). Theoretical approaches to modeling this process of speciation across an ecological gradient (e.g. Dieckmann and Doebeli 1999; Doebeli and Dieckmann 2003) have been supported by empirical evidence of ecology's role in speciation or population differentiation (Funk et al. 2005; Schluter 2009). Classical theories suggested that this

* Lighthouse Field Station, George Street, Cromarty, IV11 8YJ, UK.
• Universitetsparken 15, 2100 Copenhagen Ø, Denmark.
♦ 11 Spencer Road, Portsmouth, Hampshire PO49RN, UK.

process could only occur in allopatry, i.e. in populations with non-overlapping geographic ranges, as extrinsic barriers separating populations would be necessary to stop gene flow and lead to reproductive isolation and the formation of new species (Mayr 1942). More recent theoretical models have shown how speciation can occur across an ecological gradient in parapatry (neighboring ranges) or sympatry (overlapping ranges) (Dieckmann and Doebeli 1999). A growing number of studies have identified genetic differentiation in widespread, continuously distributed highly mobile species, in the absence of geographic barriers, e.g. the common raven *Corvus corax* Omland et al. 2000; Giraffe *Giraffa camelopardalis* Brown et al. 2007 and Grey Wolf *Canis lupus* Musiani et al. 2007. Ecological factors have been increasingly explored as an intrinsic cause of inter- and intra-specific genetic differentiation in these studies (e.g. Musiani et al. 2007; Wolf et al. 2008).

The marine habitat has few of the extrinsic barriers that may cause isolation between populations (Palumbi 1994). Additionally marine species have a relatively low cost of locomotion (Tucker 1975; Williams 1999). Many marine species therefore have large home ranges e.g. blue-fin tuna Thunnus thynnus, great white shark *Carcharodon carcharias*, and leatherback turtles *Dermochelys coriacea* (Boustany et al. 2002; Block et al. 2005; Bonfil et al. 2005; Shillinger et al. 2008). Yet despite the lack of barriers to movement, many marine species show a high degree of population structuring (e.g. Bowen et al. 1992; Pardini et al. 2001; Carlsson et al. 2004). Site fidelity to breeding or feeding grounds appears to be an important factor in determining population structure (Baker et al. 1990; Palsbøll et al. 1995; Pardini et al. 2001). However, foraging preferences can lead to differentiation across less well-defined geographic boundaries and result in resource polymorphism (Smith and Skúlason 1996). A number of well-studied whale and dolphin species appear to typify this (Hoelzel 2002).

Adaptive divergence is most rapid during the invasion of novel habitats; since the last glacial maximum and the contraction of the seasonal and permanent sea ice an abundance of new habitats have become available to whales and dolphins (Hewitt 2000). Despite the lack of geographic barriers to gene flow and the low cost of movement in the marine environment, whales and dolphins frequently show a high degree of genetic differentiation among populations. It has been hypothesized that foraging specialization may evolve after the invasion of new habitats, and there are many observed cases of foraging specialization and genetic differentiation among whale and dolphin populations, even within overlapping geographic localities. Here we review cases of ecological and genetic divergence in cetaceans and examine the micro-evolutionary processes involved. We also identify data gaps, and suggest directions for future research to increase our understanding of the role of ecology on reproductive isolation and ultimately speciation, in whales and dolphins.

Within the odontocetes, several delphinid species are known to have ecology-based polymorphisms; the bottlenose dolphin (*Tursiops truncatus*), common dolphin (*Delphinus* sp.) and spotted dolphin (*Stenella attenuata*) all have a genetically and morphologically differentiated nearshore and offshore form in at least part of their range (Douglas et al. 1984; Rosel et al. 1994; Mead and Potter 1995; Hoelzel 1998a; Kingston and Rosel 2004; Natoli et al. 2004, 2006; Tezanos-Pinto et al. 2009). Phylogenetics when coupled with character traits can provide evidence for a causal role of ecological selection on these traits as opposed to them having arisen from random drift during a period of isolation (Orr and Smith 1996). Evidence for ecological selection is particularly convincing when independent parallel divergences have taken place (Schluter and Nagel 1995). For example, Natoli et al. (2006) found that the long- and short-beaked forms of common dolphin in the Pacific had diverged independently of long- and short-beaked forms in the Atlantic indicating a role for divergent ecological selection on morphological characteristics in this case. One might speculate that the difference in beak size between the alternate morphotypes would allow for the exploitation of different prey resources, or different foraging strategies due to habitat differences. Multi-chemical dietary markers such as stable isotope or fatty acid analyses offer a promising approach to investigating this further.

Life history traits can also vary greatly and change rapidly in response to ecological variables and lead to reproductive isolation (Reznick et al. 2006; Carlson et al. 2007). The level of extrinsic mortality, in particular the level of age-dependant mortality is thought to be a key determinant of life history traits. Female short-finned pilot whales (*Globicephala macrorhynchus*) live up to 65 years, but no females over 36 were found to be pregnant (Kasuya and Marsh 1984; Marsh and Kasuya 1984; Figure 1a). In contrast even the oldest female long-finned pilot whales (*Globicephala melaena*) at 55 years can give birth (Martin and Rothery 1993; Figure 1a). This difference in life history is surprising, given the close relationship of the species, and given that both live in matrifocal groups and could contribute towards group fitness until late in life. Foote (2008) suggested that the dramatic difference in life history between the two pilot whale species was a consequence of the difference in the level of extrinsic mortality, which doubled twice as fast with increasing age in the long fin pilot whale (Figure 1b). There would therefore be much greater selection pressure on investing in somatic maintenance late in life for the short-finned pilot whale, which may explain a longer lifespan and lengthy menopause in this species. Divergence in life history traits can become linked with mate choice and lead to reproductive isolation, increasing the divergence between traits (e.g. Hendry et al. 2004). This is particularly the case when breeding is linked to a specific time due to resource availability or some other environmental factor that differs between populations. A theoretical framework of 'Isolation by Time and Adaptation by Time' suggested by Hendry and Day (2005), is likely to be applicable to a number of seasonal breeding whale species.

Figure 1a. The life history traits of three cetacean species show variation in the reproductive period of their lifespan, shown in white.

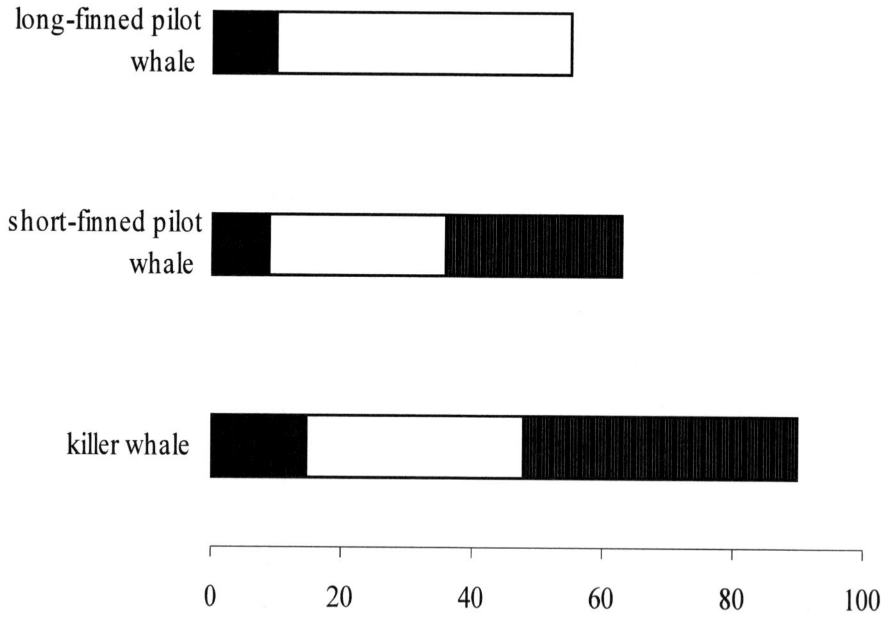

Figure 1a. The life history traits of three cetacean species show variation in the reproductive period of their lifespan, shown in white.

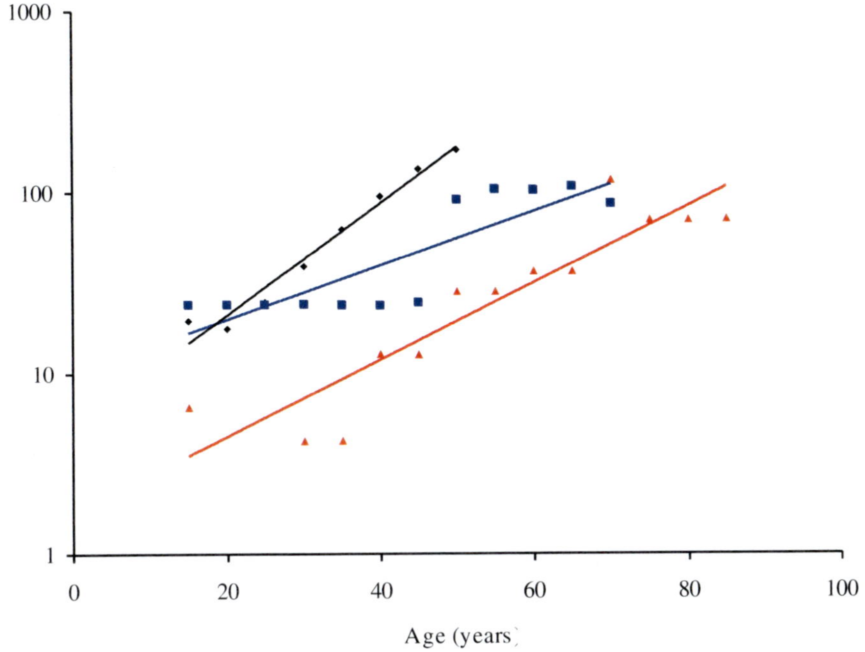

Figure 1b. The mortality-doubling rate is almost twice in long-finned pilot whales (black) of that in short-finned pilot whales (blue) or killer whales (red).

Killer whales (*Orcinus orca*) are one of the world's most widely distributed mammals, found from pole-to-pole, across all the worlds oceans and in both nearshore coastal waters and offshore pelagic waters (Rice 1998; Forney and Wade 2007). Much of this geographic range would have only been accessible since the last glacial maximum 20,000 years ago, when latitudes greater than 60° degrees were covered in seasonal or permanent sea ice (Hewitt 2000). The greatest densities of killer whales are found at higher latitudes (Forney and Wade 2007) suggesting a rapid colonization of new habitats and niches by many killer whale populations since the last glacial maximum. Larger brains are thought to be advantageous in increasing behavioral flexibility and innovative behavior, and subsequently advantageous for establishment within novel or altered environments (Sol et al. 2008). This then subsequently can lead to more rapid morphological divergence and diversification (Wyles et al. 1983; but see Lynch 1990). The killer whale has the largest brain, with the most cortical complexity, amongst the delphinids (Morino et al. 2004). A search of the literature using the databases Web of Science and Google Scholar found peer reviewed papers describing 27 distinct foraging strategies by killer whales, these included intentional standing on beaches to catch pinnipeds (Lopez and Lopez 1985; Guinet 1991), causing waves to wash seals off ice floes (Smith et al. 1981), feeding on fishing discards (Luque et al. 2007), taking fish from long-line fisheries (Yano and Dahlheim 1995) and 'carousel feeding' around herring shoals to herd them before stunning them with tail slaps (Similä and Ugarte 1993). These innovations would be key to establishing populations within new habitats and ecological niches following the last glacial maximum. Newly founded populations would then be under divergent selection from their novel environments.

Historically, taxonomists have split killer whales into several species within the genus *Orcinus* (see Heyning and Dahlheim 1988 for a review), although currently it is considered to be a monotypic single species pending further investigation (Rice 1998). However, recent research has highlighted evidence of divergence in ecology, morphology and genetics that suggest a revision in the taxonomic status of killer whales is needed (Waples and Clapham 2004; LeDuc et al. 2008). This research has focused on populations in the Northeast Pacific and Antarctic waters.

The killer whale populations of the Northeastern Pacific are the most well known, having been the subjects of over three decades of systematic research (Ford et al. 2000). Photo-identification of naturally marked individuals and analysis of both nuclear and mitochondrial DNA markers has identified several socially and reproductively isolated populations (Bigg et al. 1991; Ford et al. 2000; Hoelzel et al. 1998, 2007; Barrett-Lennard et al. 2001). Direct observation, and analyses of stomach contents and multi-chemical dietary markers (such as stable isotopes, fatty acids and persistent organic pollutants), all indicate the existence of three distinct ecotypes within the Northeastern Pacific (Ford et al. 1998; Saulitis et al. 2000; Herman et al. 2005; Krahn et al. 2007). Populations of a fish-eating ecotype, known as 'residents' due to their high site fidelity to core areas, are found all around the Pacific Rim (Forney and Wade 2007; Zerbini et al. 2007). These live in highly stable matrifocal social groups, known as pods (Bigg et al. 1991). The resident populations are found in sympatry with populations of a mammal-eating ecotype know as 'transients' (Ford and Ellis 1999). Lastly an 'offshore' ecotype which is thought to be pisciverous, and may specialize on elasmobranches (Krahn et al. 2007; Dahlheim et al. 2008) is also found in partial sympatry to both residents and transients.

There is strong evidence that each North Pacific ecotype is under divergent ecological selection, however quantitative traits measured to date showing variation between ecotypes are mostly plastic behavioral traits such as vocal behavior (Barrett-Lennard et al. 1996; Deecke et al. 2005; Foote and Nystuen 2008) or social structure (Bigg et al. 1991; Baird and Whitehead 2000). Genetically determined morphological differences appear to be slight, and have only been qualitatively reported to date (Baird and Stacey 1988). Ongoing morphometric analysis of skulls may provide the first comparison of genetically determined quantitative traits (Waples and Clapham 2004). There appears to be reproductive isolation between ecotypes; there are no shared mtDNA haplotypes between ecotypes and estimates of gene flow between populations of different sympatric ecotypes based upon nuclear DNA markers are approximately one migrant per generation (Hoelzel et al. 2007).

Three highly disparate killer whale ecotypes have also been found in Antarctic waters. Both quantitative and qualitative morphological differences are found between these Antarctic ecotypes (Pitman and Ensor 2003; Pitman et al. 2007). Type A is an open water forager that appears to specialize in hunting minke whales, its appearance is of a typical killer whale phenotype as found in the Pacific (Pitman and Ensor 2003). The type B ecotype is one of two ice-edge dwelling forms of Antarctic killer whales (Pitman and Ensor 2003; Andrews et al. 2008), it has a light grey dorsal cape and a much larger eye patch than other killer whales (Pitman and Ensor 2003). Type B killer whales mainly hunt pinnipeds by tipping them off of ice floes (Smith et al. 1981; Pitman and Ensor 2003). Lastly type C is a dwarf form of killer whale and also dwells at the ice-edge (Pitman and Ensor 2003; Andrews et al. 2008), the maximum length is 6.1 meters for this type in comparison to a maximum length of 9.2 meters for type A males (Pitman et al. 2007). Similar to type B killer whales, type C killer whales also have the dorsal cape, but they have a very small and slanted eye patch (Pitman and Ensor 2003). It is thought that they hunt fish although they have been seen harassing and possibly eating Adélie penguins (Ballard and Ainley 2005; Lauriano et al. 2007). The distinct morphology, absence of observed hybrid forms and lack of shared haplotypes between Antarctic killer whale ecotypes strongly suggests they are reproductively isolated (LeDuc et al. 2008).

Speculatively there appears to be a correlation between phenotype and resource use in these Antarctic killer whales, for example the light dorsal cape found in both ice dwelling forms could be for fixing vitamin D or camouflage within the ice floes. Similarly the large white eye patch of the type B killer whales which 'spy-hop' around ice floes looking for potential prey may offer some camouflage against the Antarctic backdrop, whereas the slim eye patch of the type C killer whale which slopes forward may be an aid to herd shoaling fish towards the mouth. Naturally it is not possible to test such hypotheses in such large mammals through classic ecological techniques such as reciprocal transplantation experiments that can be carried out on smaller study subjects (for example threespine sticklebacks *Gasterosteus* spp. Nagel and Schluter 1998). However we can look for evidence of selection on genes controlling putative ecologically selected traits (e.g. Baird et al. 2008). The advent of economical, high-throughput DNA sequencing will make this approach far quicker and more affordable than previously possible (Baird et al. 2008), and should greatly enhance our understanding of the role of divergent ecological selection on genotype and phenotype.

The evolutionary divergence of killer whales offers a useful example of the strengths and weaknesses of a phylogenetic approach to studying adaptive radiation and speciation. There is a clear phylogenetic relationship signal associated with ecology in our tree of North Pacific and Antarctic killer whale haplotypes belonging to known ecotypes (Figure 2). This would suggest that killer whales underwent an early genetic and ecological divergence. This would be consistent with theoretical models and previous empirical studies of adaptive radiation, in which a rapid colonization of available niches and evolutionary divergence due to macrohabitat differences is usually followed by slower and smaller divergence due to microhabitat differences (Schluter 1998; Gavrilets and Losos 2009). However, a drawback of such methods is that nothing is known about past character states or ecological conditions; it is not know if the ecology of killer whale lineages has been consistent over such timescales, for example when the 'transient' clade first diverged, if these ancestral 'transients' were marine mammal foraging specialists? With the ever-increasing dataset of sub-fossil killer whale samples that are becoming available for study as a result of ocean drag net fishing, multi-disciplinary studies using radiocarbon dating, stable isotope analysis and ancient DNA techniques may help resolve some of these outstanding questions on historic ecology.

Among baleen whales there is less evidence than in the odontocetes of resource polymorphism at a population level. However, recent studies have identified individuality in foraging specialization within populations in a number of baleen whale species, e.g. humpback whales (*Megaptera novaeangliae* Weinrich et al. 1992); minke whales (*Balaenoptera acutonostrata* Hoelzel et al. 1989); and right whales (genus *Eubalaena* Valenzuela et al. 2009).

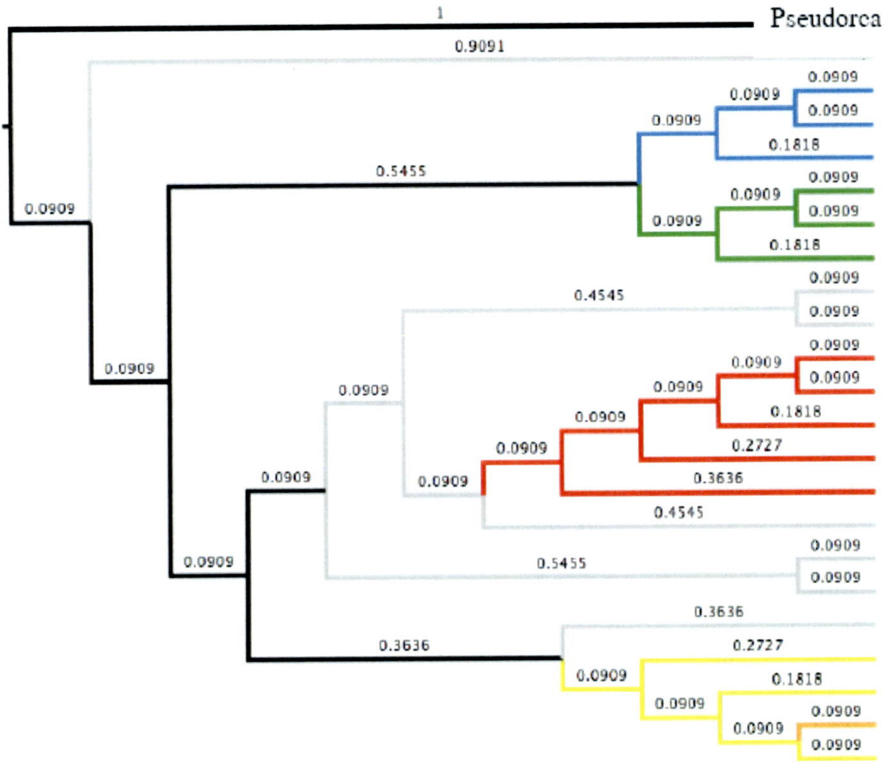

Figure 2. A neighbourhood-joining tree rooted with *Pseudorca* as an outgroup, based on sequences from LeDuc et al. (2008). Tip and node color indicate ecotype: grey=Antarctic type A; blue=Antarctic type B; green=Antarctic type C; red=Pacific transient; yellow=Pacific resident; orange=Pacific offshore.

Within-population foraging specialization has been found in right whales as a result of stable isotope studies on calving grounds. Southern right whales (*E. australis*) populations are scattered across the southern hemisphere, with ten coastal winter calving grounds currently recognized by the International Whaling Commission (IWC 2001). Genetic studies have confirmed that southern right whales show strong maternal fidelity to winter calving grounds, with these grounds showing significant differentiation of mitochondrial haplotypes across the southern hemisphere (Patenaude et al. 2007). Limited genetic data from right whale summer feeding areas in south Georgia and off western Australia suggests that these areas contain a mixture of animals from adjacent calving grounds (Patenaude et al. 2007).

During the summer, right whales move offshore to feed on krill and copepod species, and are seen at a range of latitudes (Townsend 1935), traveling thousands of kilometers to forage (Best et al. 1993). Evidence from stable isotope studies in Peninsula Valdez (Argentina) suggests that right whales sharing common winter calving grounds have diverse summer foraging strategies, which are maternally inherited (Valenzuela et al. 2009). Stable isotopes of carbon and nitrogen vary according to latitude in the ocean, as phytoplankton take up carbon dioxide at rates dependent on sea temperature and proximity to the coast. Stable isotopes in whale skin are therefore informative about the latitudinal range of the population during summer foraging. Among Argentinean right whales, skin samples revealed a broad range of isotope values, with individuals of the same mitochondrial haplotype having more similar isotope values than those of different haplotypes. The broad range of isotopic values is consistent with whaling records from the region, which found that whales killed south of 50°S had stomachs filled with krill, those north of 40°S had stomachs filled with copepods and those killed between 40-50°S had stomachs filled with both copepods and krill (Tormosov et al. 1998). Using a randomization test, Valenzuela et al. (2009) found that the isotopic similarity among matrilines was strongly significant, indicating that site fidelity to feeding grounds is inherited over at least several generations within the calving ground.

Interestingly, some matrilines had relatively large stable isotope ranges, suggesting that individuals within those groups show some plasticity in foraging locations and prey types. This finding is consistent with a recent analysis of lifetime foraging 'paths' using stable isotope values from baleen plates of stranded whales in Peninsula Valdez (Rowntree et al. 2008). Among the five animals surveyed, two showed narrow stable isotope values, indicating fidelity to specific geographic feeding areas, while the other three showed significantly broader ranges, suggesting a more flexible lifetime foraging strategy.

Valenzuela et al. (2009) conclude that strong fidelity in right whale foraging strategies may be a major cause of the slow apparent recovery from exploitation for some populations, as it limits their ability to access new feeding opportunities and renders them vulnerable to local changes in productivity. The effects of fluctuations in prey abundance are likely to be particularly pronounced in regions where populations are small, such as the western north Atlantic (Kenney et al. 2001). Correlations between reproductive success (calving rates) and local sea surface temperature have been found in north Atlantic (Greene et al. 2001) which support this idea. A study of calving rates during El Nino years also show a similar pattern in the south Atlantic (Leaper et al. 2006), although this population is much larger and may therefore be anticipated to be more robust to local temperature anomalies.

Most humpback whale populations worldwide are highly migratory, spending winter months engaged in breeding and calving activity in low-latitude tropical waters and summer months feeding at high latitudes. The species exhibits strong matrilineal structuring on breeding grounds, with frequencies of maternal haplotype (mitochondrial DNA) lineages in each population differing significantly from those of adjoining breeding grounds across both hemispheres (Baker et al. 1993; Baker and Medrano-González 2002). This is thought to occur as a result of the early maternal migratory experience of calves (Baker et al. 1990). Maternal population differentiation between feeding areas is less marked, although clines in haplotype frequencies across feeding areas have been found in both north Atlantic (Palsbøll et al. 1995; Larsen et al. 1996) and north Pacific oceans (Baker and Medrano-González 2002), suggesting some stratification on feeding grounds also. Strong maternal fidelity to natal breeding grounds may be the driving mechanism for the pronounced female population structure found between humpback stocks; however animals from multiple breeding stocks are known to mix and associate on shared feeding grounds (e.g. Calambokidis et al. 2008).

Humpbacks are considered the most diverse foragers among all baleen whales. They feed mainly on large zooplankton (such as krill) and a variety of small schooling fish, either alone or in small groups (usually 2-3 members). Male-male group associations are less common than those among mature females (without calves) or between females and males (Weinrich and Kuhlberg 1991). Most associations are relatively brief in duration. Foraging strategies include the use of clouds, columns or nets of exhaled bubbles to concentrate krill or fish ("bubble-net" feeding), the herding (and possibly disabling) of prey by flicking or pounding with the flukes and flippers, and feeding in formation ("echelon" feeding). While it has been hypothesized that feeding associations made during group foraging may be a more efficient means of capturing patchily distributed prey of fluctuating density (e.g. krill) or maximizing capture of schooling fish (e.g. herring), it is still uncertain whether groups are operating cooperatively, or whether individuals are behaving in a commensalistic or parasitic manner (see Weinrich et al. 2009).

While in the past it has been hypothesized that kin selection might drive foraging associations among humpbacks (e.g. Baker 1985; Weinrich 1991; Baker et al. 1994), no evidence for first-order kin associations within foraging groups have so far been found. A long-term photo-identification and genetic study in southeast Alaska found that while 'core' individuals within herring specialist foraging groups showed consistent associations over a number of years, microsatellite analyses (41 individuals) suggested that they were not more closely related to their foraging group than to the feeding population at large (Sharpe 2001). Parallel photo-identification of individuals revealed that calves born to 'core' herring foragers rarely joined these groups as adults, while behavioural observations suggested that calves do not actively participate in this style of foraging while still associated with the mother. This accords with the observation that animals learn foraging strategies from peers after weaning, around 1-2 years of age (Weinrich et al. 1992). There may also be a greater benefit to flexibility in associations among foraging humpbacks since many different combinations of behaviours and tactics may be required to maximize consumption of prey which vary, sometimes unpredictably, in geographic location, density, and group dynamics (Clapham 1996). Long-term studies in the Gulf of Maine have witnessed a steady increase in a novel feeding behaviour, termed 'lob-tailing', since 1980, from 0.0% to 50.6% between 1980 and 1989 (Weinrich et al. 1992). This behaviour combines bubble-net feeding with an initial tail slap on the surface, and is used in foraging for sand lance. It has been hypothesized that the behaviour has arisen in concert with a shift in prey species, since the regional herring stocks declined rapidly in the 1970s due to overfishing. Despite the substantial increase in this behaviour among individuals over time, an associated long-term photo-identification study of mothers and calves in the region revealed that the novel feeding technique was often taken up by the calf after weaning, but rarely by the mother. This study provides further support for the hypothesis that feeding behaviour is not maternally transmitted (Weinrich et al. 1992). While behavioural theory predicts that animals participating in activities which provide a group benefit, such as foraging, are more likely to be related (Hamilton 1964a; Hamilton 1964b), the promiscuous breeding system of humpbacks (multiple male suitors for each female) means that animals possess much weaker kinship ties across cohorts than among other social foraging mammals such as elephants and wolves. At present there is no evidence to support kin foraging among humpback whales, although a wider study of genetic relatedness among foraging groups is needed. Current evidence instead suggests that non-related foragers can have stable feeding ground associations in the long term, for reasons not yet fully understood (Sharpe 2001; Weinrich et al. In review).

Interestingly, at a population level, a pattern of feeding group matrilineal (mitochondrial DNA) associations has been detected in the north Atlantic Gulf of Maine (Weinrich et al. 2006), over all foraging groups and also among female-only groups. This finding therefore suggests that while foraging strategies may vary among related individuals, humpback prey preferences on a larger scale may be heritable, or maternally influenced during the first year of life. An example of this was seen in 1987-1988 in the North Atlantic, when 10 mackerel-feeding humpbacks died of dinoflagellate poisoning in Cape Cod, Massachusetts. This group showed previous evidence of association, and many members were known summer visitors from local sightings histories in the southern Gulf of Maine. The group genetic mitochondrial DNA composition was significantly different from that of the larger southern Gulf of Maine population, which feeds on sand lance as the major prey. The majority of the mackerel feeders bore two mitochondrial matrilines, one of which was rare and the other non-existent in the larger population. The group had also been documented feeding in prior months at Jeffrey's Ledge, a region where herring (another mid-water pelagic species) are the main prey. There is therefore some evidence suggestive of more indirect genetic preferences for certain prey types among foraging humpbacks (Baker et al. 1994; Weinrich et al. 2006), although the precise mechanism for this relationship remains unclear. Extensive further study of the genetic and behavioural relationships among individuals within cooperative pods exercising different foraging strategies may in the long term shed more light on this interesting question.

The relationship between ecology, behavioral plasticity and evolutionary divergence is a complex one. Behavioral plasticity can on the one hand drive rapid evolutionary divergence by facilitating the exploitation of new habitats or resources. However, behavioral plasticity can also act as a buffer during periods of environmental change and stabilize any change in genetically determined phenotypic traits (Duckworth 2009). Whether behavioral plasticity acts as a catalyst or buffer to evolutionary divergence depends upon whether a behavioral shift is in response to environmental change or if it allows an animal to occupy a new environment. This distinction is also a key determinant of the mode of speciation (Duckworth 2009). In cases where the environment changes rapidly and the animal adapts to this through a behavioral shift then evolutionary divergence is likely to be relatively slow, but the animal will still be subject to some novel selection pressures due to the change in its interaction with the altered environment. Such a scenario could lead to micro-evolutionary changes that ultimately can lead to sympatric speciation (Duckworth 2009). In cases where behavioral plasticity allows some animals to move to a new environment where they are subjected to new selection pressures, evolutionary divergence should be relatively rapid and is most likely to be either allopatric or parapatric.

The population level differences in morphology, ecology and genetics found in the odontocetes may be a result of divergent selection pressures following the exploitation of new environments; whereas the within-population variation found in baleen whales may be determined by past shifts in behavior during a period of rapid environmental change in existing habitats. The degree of these differences is also likely to be dependent upon the proportion of individuals within a population experiencing new environmental selection pressures. For selection from any new environmental variable to act upon genetic traits and be fixed within the population depends upon a high proportion of the individuals within the populations being subjected to the variable over an evolutionary timescale.

The annual migratory cycle and strong site fidelity to feeding and breeding grounds in humpback and right whales may restrict their ability to move to new habitats in response to environmental change. They may instead respond behaviorally to environmental change as found by Weinrich et al (1992) for humpback whales in the Gulf of Maine. As noted above, this is likely to lead to slower, sympatric divergence due to micro-evolutionary processes resulting from differences between individuals in their interaction with the environment. Additionally, the fluid social structure of baleen whales has led to individuality in foraging methods. These new foraging strategies would need to be taken up by a large proportion of the population to lead to genetic fixation of any traits under selection from the environment.

In odontocetes, the social structure tends to have developed towards more cohesive and extended kin-based groups (e.g. Bigg et al. 1990), and foraging methods are shared throughout the population (e.g. Ford et al. 1998; Herman et al. 2005). This would enable traits under selection due to this interaction with the environment to become fixed at a population level. Additionally, although many odontocetes populations are site faithful based on the presence of a predictable resource, there is also strong evidence, as noted above for killer whales, that ranges were recently expanded- probably due to behavioral shifts. This would have led to a more rapid evolutionary divergence between populations, probably in parapatry and allopatry.

In summary, behavioral plasticity in many whale and dolphin species has allowed variation at both the population and individual level in interactions with the environment. This variation in ecology is reflected in the divergence in morphology and genetics due to the different selection pressures that arise from divergent ecological interactions. The literature on ecological speciation is dominated by studies of model species for which classic experimental techniques can be used to identify the mechanisms and processes involved in evolutionary divergence. The advent of new molecular techniques in combination with field observations promises to allow a much greater understanding of these mechanisms and processes in non-model organisms such as whales and dolphins. We look forward to seeing rapid progress in this field in the coming decade.

REFERENCES

Andrews, RD; Pitman, RL; Ballance, LT. Satellite tracking reveals distinct movement patterns for Type B and Type C killer whales in the southern Ross Sea, Antarctica. *Polar Biology*, 2008 31, 1461-1468.

Baird, NA; Etter, PD; Atwood, TS; Currey, MC; Shiver, AL; Lewis, ZA; Selker, EU; Cresko, WA; Johnson, EA. Rapid SNP discovery and genetic mapping using sequenced RAD markers. *PLoS One*, 2008 10: e3376.

Baird, RW; Stacey PJ. Variation in saddle patch pigmentation in populations of killer whales (*Orcinus orca*) from British Columbia, Alaska, and Washington State. *Canadian Journal of Zoology*, 1988 66, 2582-2585.

Baird, RW; Whitehead, H. Social organization of mammal-eating killer whales: group stability and dispersal patterns. *Canadian Journal of Zoology*, 2000 78, 2096-2105.

Baker, CS. The population structure and social organization of humpback whales (*Megaptera novaeangliae*) in the Central and Eastern North Pacific. PhD thesis: University of Hawaii. Honolulu; 1985

Baker, CS; Medrano-González, L. Worldwide distribution and diversity of humpback whale mitochondrial DNA lineages. In: Pfeiffer CJ, editor. *Molecular and Cell Biology of Marine Mammals*. Malabar: Krieger Publishing Company; 2002; 84-99.

Baker, CS; Palumbi, SR; Lambertsen, RH; Weinrich, MT; Calambokidis, J; O'Brien, SJ. Influence of seasonal migration on the distribution of mitochondrial DNA haplotypes in humpback whales. *Nature* 1990 344, 238-240.

Baker, CS; Perry, A; Bannister, JL; Weinrich, MT; Abernethy, RB; Calambokidis, J; Lien, J; Lambertsen, RH; Urbán Ramírez, J; Vasquez, O; Clapham, PJ; Alling, A; O'Brien, SJ; Palumbi, SR. Abundant mitochondrial DNA variation and world-wide population structure in humpback whales. *Proceedings of the National Academy of Sciences of the United States of America*, 1993 90, 8239-8243.

Baker, CS; Weinrich, MT; Early, G; Palumbi, SR. Genetic impact of an unusual Group mortality among humpback whales. *Journal of Heredity* 1994 85, 52-54.

Ballard, G; Ainley, DG. Killer whale harassment of Adélie penguins at Ross Island. *Antarctic Science*, 2005 17, 385-386.

Barrett-Lennard, LG; Ford, JKB; Heise, KA. The mixed blessing of echolocation: differences in sonar use by fish-eating and mammal-eating killer whales. *Animal Behaviour*, 1996 51, 553-565.

Barrett-Lennard, LG; Ellis, GM. Population structure and genetic variability in Northeastern-Pacific killer whales: Towards an assessment of population viability. *Canadian Science Advisory Secretariat Research Document*, 2001 065.

Best, PB; Payne, R; Rowntree, VJ; Palazzo, JT; Both, MD. Long-range movements of South Atlantic right whales (*Eubalaena australis*). *Marine Mammal Science*, 1993 9, 227-234.

Bigg, MA; Olesiuk, PF; Ellis, GM; Ford, JKB; Balcomb III, KC. Social organization and genealogy of resident killer whales (*Orcinus orca*) in the coastal waters of British Columbia and Washington State. *Reports of the International Whaling Commission special issue*, 1990 12, 383-405.

Block, BA; Teo, SLH; Walli, A; Boustany, A; Stokesbury, MJW; Farwell, CJ; Weng, KC; Dewar, H; Williams, TD. Electronic tagging and population structure of Atlantic bluefin tuna. *Nature*, 2005 434, 1121-1127.

Bonfil, R; Meyer, M; Scholl, MC; Johnson, R; O'Brien, S; Oosthuizen, H; Swanson, S; Kotze, D; Paterson, M. Transoceanic migration, spatial dynamics, and population linkages of white sharks. *Science*, 2005 310, 100-103.

Boustany, AM; Davis, SF; Pyle, P; Anderson, SD; Le Boeuf, BJ; Block, BA. Expanded niche for white sharks. *Nature*, 2002 415, 35-36.

Bowen, BW; Meylan, AB; Ross, JP; Limpus, CJ; Balazs, GH; Avise, JC. Global population structure and natural history of the green turtle (*Chelonia mydas*) in terms of matriarchal phylogeny. *Evolution*, 1992 46, 865-881.

Brown, DM; Brenneman, RA; Koepfli, K-P; Pollinger, JP; Mila, B; Georgiadis, NJ; Lois Jr, EE; Grether, GF; Jacobs, DK; Wayne, RK. Extensive population genetic structure in the giraffe. *BMC Biology*, 2007 5:57.

Calambokidis, J; Falcone, EA; Quinn, TJ; Burdin, AM; Clapham, PJ; Ford, JKB; Gabriele, CM; LeDuc, R; Mattila, DK; Rojas-Bracho, L; Straley, JM; Taylor, BL; Urbán-Ramirez, J; Weller, RD; Witteveen, BH; Yamaguchi, M; Bendlin, A; Camacho D; Flynn, K; Havron, A; Huggins, J; Maloney, N. *SPLASH: Structure of populations, levels of abundance and status of humpback whales in the North Pacific.* Cascadia Research. 2008.

Carlson, SM; Hilborn, R; Hendry, AP; Quinn, TP. Predation by bears drives senescence in natural populations of salmon. *PLoS ONE*, 2007 2: 1286.

Carlsson, J; McDowell, JR; Diaz-Jaimes, P; Carlsson, JEL; Boles, SR; Gold JR; Graves, JE. Microsatellite and mitochondrial DNA analyses of Atlantic bluefin tuna (*Thunnus thynnus thynnus*) population structure in the Mediterranean Sea. *Molecular Ecology*, 2004 13, 3345-3356.

Clapham, PJ. The social and reproductive biology of humpback whales: an ecological perspective. *Mammal Review* 1996 26, 27-49.

Dahlheim, ME; Schulman-Janiger, A; Black, N; Ternullo, R; Ellifrit, D; Balcomb III, KC. Eastern temperate North Pacific offshore killer whales (*Orcinus orca*): Occurrence, movements, and insights into feeding ecology. *Marine Mammal Science*, 2008 24: 719-729.

Deecke, VB; Ford, JKB; Slater, PJB. The vocal behaviour of mammal-eating killer whales: communicating with costly calls *Animal Behaviour*, 2005 69, 395-405.

Dieckmann, U; Doebeli, M. On the origin of species by sympatric speciation. *Nature*, 1999 400, 354–357.

Dobzhansky, T. Genetics and the Origin of Species. 3rd edn. New York: Columbia University Press; 1951.

Doebeli, M; Dieckmann, U. Speciation along environmental gradients. *Nature*, 2003 421, 259–264.

Douglas, ME; Schnell, GD; Hough, DJ. Differentiation between inshore and offshore spotted dolphins in the eastern tropical Pacific Ocean. *Journal of Mammalogy*, 1984. 65, 375-387.

Duckworth, RA. The role of behavior in evolution: a search for mechanism. *Evolutionary Ecology*, 2009 23, 513-531.

Endler, J. Geographic Variation, Speciation and Clines. Princeton University Press; 1977.

Foote, AD. Mortality rate acceleration and post-reproductive lifespan in matrilineal whale species. *Biology Letters*, 2008 4, 189-191.

Foote, AD; Nystuen, JA. Variation in call pitch among killer whale ecotypes. *Journal of the Acoustical Society of America*, 2008 123, 1747-1752.

Ford, JKB; Ellis, GM; Barrett-Lennard, LG; Morton, AB; Palm, RS; Balcomb, KC. Dietary specialization in two sympatric populations of killer whale (*Orcinus orca*) in coastal British Columbia and adjacent waters. *Canadian Journal of Zoology*, 1998 76, 1456–1471.

Ford, JKB; Ellis, GM. Transients: Mammal-Hunting Killer Whales of British Columbia, Washington, and Southeastern Alaska. Vancouver: UBC Press; 1999.

Ford, JKB; Ellis, GM; Balcomb, KC. Killer Whales; The Natural History and Genealogy of *Orcinus orca* in British Columbia and Washington. Vancouver: UBC Press; 2000.

Forney, KA; Wade, P. Worldwide distribution and abundance of killer whales. In: Estes J, editor. *Whales, Whaling and Ecosystems*. University of California Press; 2007; 145–162.

Funk, DJ; Nosil, P; Etges, WJ. Ecological divergence exhibits consistently positive associations with reproductive isolation across disparate taxa. *Proceedings of the National of Academy of Sciences, USA*, 2006 103, 3209–3213.

Gavrilets, S; Losos, JB. Adaptive radiation: Contrasting theory with data. *Science*, 2009 323, 732–737.

Greene, CH; Pershing, AJ; Kenney, RD; Jossi. JW. Impact of climate variability on the recovery of endangered north Atlantic right whales. *Oceanography*, 2001 16. 98-103.

Guinet, C. Intentional stranding apprenticeship and social play in killer whales (*Orcinus orca*). *Canadian Journal of Zoology*, 1991 69, 2712-2716.

Hamilton, WD. The genetical evolution of social behaviour. I. *Journal of Theoretical Biology* 1964a. 7, 1-16.

Hamilton, WD. The genetical evolution of social behaviour. II. *Journal of Theoretical Biology*, 1964b 7, 17-52.

Hendry, AP; Morbey, YE; Berg, OK; Wenburg, JK. Adaptive variation in senescence: reproductive lifespan in a wild salmon population. *Proceedings of the Royal Society of London series B*, 2004 271, 259-266.

Hendry, AP; Day, T. Population structure attributable to reproductive time: isolation by time and adaptation by time. *Molecular Ecology*, 2005 14, 901–916.

Herman, DP; Burrows, DG; Wade, PR; Durban, JW; Matkin, CO; LeDuc, RG; Barrett-Lennard, LG; Krahn, MM. Feeding ecology of eastern North Pacific killer whales *Orcinus orca* from fatty acid, stable isotope and organochlorine analyses of blubber biopsies. *Marine Ecology Progress Series*, 2005 302, 275-291.

Hewitt, G. The genetic legacy of the Quaternary ice ages. *Nature*, 2000 405, 907-913.

Heyning, JE; Dahlheim, ME. *Orcinus orca. Mammalian Species*, 1988 304. 1-9

Hoelzel, AR; Dorsey, EM; Stern, SJ. The foraging specializations of individual minke whales. *Animal Behaviour*, 1989 38, 786-794.

Hoelzel, AR; Potter, CW; Best, PB. Genetic differentiation between parapatric "nearshore" and "offshore" populations of the bottlenose dolphin. *Proceedings of the Royal Society London series B*, 1998a 265, 1177-1183.

Hoelzel, AR; Dahlheim, ME; Stern, SJ. Low genetic variation among killer whales in the eastern North Pacific, and genetic differentiation between foraging specialists. *Journal of Heredity* 1998b 89, 121-128.

Hoelzel, AR. Resource specialization and the evolution of population genetic structure in delphinid species In: CJ Pfeiffer, editor. *Molecular and cell biology of marine mammals.* Malabar, Florida: Krieger Publishing; 2002.

Hoelzel, AR; Hey, J; Dahlheim, ME; Nicholson, C; Burkanov, V; Black, N. Evolution of population structure in a highly social top predator, the killer whale. *Molecular Biology and Evolution*, 2007 24, 1407-1415.

IWC. Report of the workshop on the comprehensive assessment of right whales: a worldwide comparison. *Journal of Cetacean Research and Management Special Issue*, 2001 2, 1-60.

Kasuya, T; Marsh, H. Life history and reproductive biology of the short-finned pilot whale, *Globicephala macrorhynchus*, off the Pacific Coast Japan. Report. *Reports of the International Whaling Commission Special Issue*, 1984 6, 259-310.

Kenney, RD; Mayo, CA; Winn, HE. Migration and foraging strategies at varying spatial scales in western North Atlantic right whales: a review of hypotheses. *Journal of Cetacean Research and Management Special Issue*, 2001 2, 251-260.

Kingston, S; Rosel, PE. Determining genetic differentiation among recently diverged delphinid taxa using amplified fragment length polymorphism markers. *Journal of Heredity*, 2004 95, 1-10.

Krahn, MM; Herman, DP; Matkin, CO; Durban, JW; Barrett-Lennard, L; Burrows, DG; Dahlheim, ME; Black, N; Leduc, RG; Wade, PR. Use of chemical tracers in assessing the diet and foraging regions of eastern North Pacific killer whales. *Marine Environmental Research*, 2007 63, 91-114.

Krahn, MM; Pitman, RL; Burrows, DG; Herman, DP; Pearce, RW. Use of chemical tracers to assess diet and persistent organic pollutants in

Antarctic Type C killer whales. *Marine Mammal Science*, 2008 24, 643-663.

Larsen, AH; Sigurjonsson, J; Øien, N; Vikingsson, G; Palsbøll, P. Populations genetic analysis of nuclear and mitochondrial loci in skin biopsies collected from central and northeastern North Atlantic humpback whaes (Megaptera novaeangliae): Population identity and migratory destinations. *Proceedings of the Royal Society of London Series B*, 1996 263, 1611-1618.

Lauriano, G; Fortuna, CM; Vacchi, M. Observation of killer whale (*Orcinus orca*) possibly eating penguins in Terra Nova Bay, Antarctica. *Antarctic Science*, 2007 19, 95-96

Leaper, R; Cooke, J; Trathan, P; Reid, K; Rowntree, V; Payne, R. Global climate drives southern right whale (*Eubalaena australis*) population dynamics. *Biology Letters*, 2006 2, 289-292.

LeDuc, RG; Robertson, KM; Pitman, RL. Mitochondrial sequence divergence among Antarctic killer whale ecotypes is consistent with multiple species. *Biology Letters*, 2008 4, 426-429.

Lopez, JC; Lopez D. Killer whales (*Orcinus orca*) of Patagonia, and their behavior of intentional stranding while hunting nearshore. *Journal of Mammalogy*, 1985 66, 181-183.

Luque, PL; Davis, CG; Reid, DG; Wang, J; Pierce GJ. Opportunistic sightings of killer whales from Scottish pelagic trawlers fishing for mackerel and herring off North Scotland (UK) between 2000 and 2006. *Aquatic Living Resources*, 2006 19, 403-410

Lynch, M. The rate of morphological evolution in mammals from the standpoint of the neutral expectation. *American Naturalist*, 1990 136, 727-741.

Marsh, H; Kasuya, T. Ovarian changes in the short-finned pilot whale, *Globicephala macrorhynchus. Reports of the International Whaling Commission Special Issue*, 1984 6, 311-335.

Martin, AR; Rothery, P. Reproductive parameters of female long-finned pilot whales (*Globicephala melas*) around the Faroe Islands. *Reports of the International Whaling Commission Special Issue*, 1993 14; 263-304.

Mayr, E. *Systematics and the Origin of Species*. New York: Columbia University Press; 1942.

Mead, JG; Potter, CW. Recognizing two populations of the bottlenose dolphin (*Tursiops truncatus*) off the Atlantic coast of North America; morphologic and ecologic consideration. *IBI Reports*, 1995 5, 31-44.

Morino, L; Sherwood, CC; Delman, BN; Tang, CY; Naidich, TP; Hof, P. Neuroanatomy of the killer whale (*Orcinus orca*) from magnetic resonance images. *The Anatomical Record Part A*, 2004 281A, 1256-1263.

Musiani, M; Leonard, JA; Cluff, HD; Gates, CC; Mariani, S; Paquet, PC; Vilàs, C; Wayne, RK. Differentiation of tundra/taiga and boreal coniferous forest wolves: genetics, coat colour and association with migratory caribou. *Molecular Ecology*, 2007 16, 4149-4170.

Nagel, L; Schluter, D. Body size, natural selection, and speciation in sticklebacks. *Evolution*, 1998 51, 1469-1480.

Natoli, A; Peddemors, VM; Hoelzel, AR. Population structure and speciation in the genus Tursiops based on microsatellite and mitochondrial DNA analyses. *Journal of Evolutionary Biology*, 2004 17, 363-375.

Natoli, A; Canadas, A; Peddemors, VM; Aguilar, A; Vaquero, C; Fernández-Piqueras, P; Hoelzel, AR. Phylogeography and alpha taxonomy of the common dolphin (*Delphinus* sp.). *Journal of Evolutionary Biology*, 2006 19, 943-954.

Omland, KE; Tarr, CL; Boarman, WI; Marzluff, JM; Fleischer, RC. Cryptic genetic variation and paraphyly in ravens. *Proceedings of the Royal Society of London series B*, 2000 267, 2475-2482.

Orr, MR; Smith, TB. Ecology and speciation. *Trends in Ecology and Evolution,* 1998 13, 502-506.

Pardini, AT; Jones, CS; Noble, LR; Kreiser, B; Malcolm, H; Bruce, BD; Stevens, JD; Cliff, G; Scholl, MC; Francis, M; Duffy, CAJ; Martin, AP. Sex-biased dispersal of great white sharks. *Nature* 2001 412, 139-140.

Palsbøll, P; Clapham, PJ; Mattila, DK; Larsen, F; Sears, R; Siegismund, HR; Sigurjonsson, J; Vasquez, O; Arctander, P. Distribution of mtDNA haplotypes in North Atlantic humpback whales: the influence of behaviour on population structure. *Marine Ecology Progress Series*, 1995 116, 1-10.

Palumbi, SR. Genetic divergence, reproductive isolation, and marine speciation. *Annual Review of Ecology and Systematics* 1994 25, 547-572.

Patenaude, N J; Portway, VA; Schaeff, CM; Bannister, JL; Best, P; Payne, RS; Rowntree, VJ; Rivarola, M; Baker, CS.. Mitochondrial DNA diversity and population structure among southern right whales (*Eubalaena australis*). *Journal of Heredity*, 2007 98, 147-157.

Pitman, RL; Ensor, P. Three forms of killer whales (*Orcinus orca*) in Antarctic waters. *Journal of Cetacean Research Management*, 2003 5, 131-139.

Pitman, RL; Perryman, WL; LeRoi, D; Eilers, E. A dwarf form of killer whale in Antarctica. *Journal of Mammalogy*, 2007 88, 43–48.

Reznick, D; Bryant, M; Holmes, D. The evolution of senescence and post-reproductive lifespan in guppies (*Poecilia reticulata*). *PLOS Biology*, 2006 4, 136-143.

Rice, DW. *Marine mammals of the world: Systematics and distribution*. Lawrence, Kansas: Allen Press; 1998.

Rosel, PE; Dizon, AE; Heyning, JE. Genetic analysis of sympatric populations of common dolphins (genus *Delphinus*). *Marine Biology*, 1994 119, 159-167.

Rowntree, VJ; Valenzuela, LO; Franco Fraguas, P; Seger, J. Foraging behaviour of southern right whales (*Eubalaena australis*) inferred from variation of carbon stable isotope ratios in their baleen. Paper SC/60/BRG23 presented to the IWC Scientific Committee, June 2008 (unpublished). 10pp. [Available from the office of this Journal].

Saulitis, EL; Matkin, CO; Barrett-Lennard, LG; Heise, KA; Ellis, GM. Foraging strategies of sympatric killer whale (*Orcinus orca*) populations in Prince William Sound, Alaska. *Marine Mammal Science*, 2000 16, 94-109.

Schluter, D. Ecological causes of speciation. In: Howard DJ; Berlocher SH, editors. *Endless forms: species and speciation*. Oxford University Press; 1998; 114-129.

Schluter, D. 2009. Evidence for ecological speciation and its alternative. *Science*, 323, 737-741.

Schluter, D; Nagel, LM. Parallel speciation by natural selection. *American Naturalist* 1995 146, 292-301.

Sharpe, FA. Social foraging of the Southeast Alaskan humpback whale, *Megaptera novaeangliae*. PhD Thesis: Simon Fraser University; 2001.

Shillinger, GL; Palacios, DM; Bailey, H; Bograd, SJ; Swithenbank, AM; Gaspar, P; Wallace, BP; Spotila, JR; Paladino, FV; Piedra, R; Eckert, SA; Block, BA. Persistent leatherback turtle migrations present opportunities for conservation. *PLoS Biology* 2008 6, 1408-1416.

Similä, T; Ugarte, F. Surface and under-water observations of cooperatively feeding killer whales in northern Norway. *Canadian Journal of Zoology*, 1993 71, 1494-1499.

Smith, TG; Siniff, DB; Reichle, R; Stone, S. Coordinated behavior of killer whales, *Orcinus orca*, hunting a crab-eater seal, *Lobodon carcinophagus*. *Canadian Journal Of Zoology*, 1981 59, 1185-1189.

Smith, TB; Skúlason, S. Evolutionary significance of resource polymorphisms in fishes, amphibians, and birds. *Annual Review of Ecology and Systematics*, 1996 27, 111-133.

Sol, D; Bacher, S; Reader, SM; Lefebvre, L. Brain size predicts the success of mammal species introduced into novel environments. *The American Naturalist*, 2008 172, S63-S71.

Tezanos-Pinto, G; Baker, CS; Russell, K; Martien, KK; Baird, R; Hutt, WA; Stone, G; Mignucci-Giannoni, AA; Caballero, S; Endo, T; Lavery, S; Oremus, M; Olavarría C; Garrigue, C. A worldwide perspective on the population structure and genetic diversity of bottlenose dolphins (*Tursiops truncatus*) in New Zealand. *Journal of Heredity*, 2009 100, 11-24.

Tormosov, DD; Mikhaliev, YA; Best, PB; Zemsky, VA; Sekiguchi, K; Brownell Jr. RL. Soviet catches of Southern Right Whales *Eubalaena australis*, 1951-1971. Biological data and conservation implications. *Biological Conservation*, 1998 86, 185-197.

Townsend, CH. The distribution of certain whales as shown by logbook records of American whaleships. *Zoologica N.Y.*, 1935 19, 1-50.

Tucker, VA. The energetic cost of moving about. *American Scientist*, 1975 63, 413-419.

Valenzuela, LO; Sironi, M; Rowntree, VJ; Seger, J. Isotopic and genetic evidence for culturally inherited site fidelity to feeding grounds in southern right whales (*Eubalaena australis*). *Molecular Ecology*, 2009 18, 782-791.

Waples, RS; Clapham, PJ. Report of the working group on killer whales as a case study. Report of the workshop on shortcomings of cetacean taxonomy, 28-29 April 2004, La Jolla, California.

Weinrich, MT. Long term stability in grouping patterns of humpback whales (*Megaptera novaeangliae*) in the southern Gulf of Maine. *Canadian Journal of Zoology*, 1991 69, 3012-3019.

Weinrich, MT; Rosenbaum, HC; Baker, CS; Blackmer, AL; Whitehead, H. The Influence of maternal lineages on social affiliations among humpback whales (*Megaptera novaeangliae*) on their feeding grounds in the Southern Gulf of Maine. *Journal of Heredity*, 2006 97, 226-234.

Weinrich, MT; Schilling, MR; Belt, CR. Evidence for acquisition of a novel feeding behavior: lobtail feeing in humpback whales, *Megaptera novaeangliae*. *Animal Behaviour*, 1992 44, 1059-1072.

Weinrich, MT; Steel, D; Baker, CS. In review. Neither consort nor kin: Molecular Ecology of a long-term association between feeding humpback whales (*Megaptera novaeangliae*). *Marine Mammal Science*.

Weinrich, M; Cholewiak, D; Friedlander, A; Hurst, T; Thompson, M; Ware, C; Wiley, D. Underwater behavior of associated humpback whales in the Southern Gulf of Maine. Abstracts for the 18th Biennial Conference on the Biology of Marine Mammals, Québec, Canada, 28 October 12-16, 2009.

Williams, TM. The evolution of cost efficient swimming in marine mammals: limits to energetic optimization. *Philosophical Transactions of the Royal Society B*, 1999 354, 193-201.

Wolf, JBW; Harrod, C; Brunner, S; Salazar, S; Trillmich, F; Tautz, D. Tracing early stages of species differentiation: Ecological, morphological and genetic divergence of Galápagos sea lion populations. *BMC Evolutionary Biology*, 2008 8: 150.

Wyles, JS; Kunkel, JG; Wilson, AC. Birds, behavior, and anatomical evolution. *Proceedings of the National Academy of Science USA*, 1983 80, 4394-4397.

Yano, K; Dahlheim, ME. Killer whale, orcinus-orca, depredation on longline catches of bottomfish in the southeastern bering sea and adjacent waters. *Fishery Bulletin*, 1995 93, 355-372.

Zerbini, AN; Waite, JM; Durban, JW; LeDuc, R; Dahlheim, ME; Wade, PR. Estimating abundance of killer whales in the nearshore waters of the Gulf of Alaska and Aleutian Islands using line-transect sampling. *Marine Biology*, 2007 150, 1033-1045.

In: Whales and Dolphins
Editor: Craig A. Murray, pp. 113-125
ISBN 978-1-61668-693-2

Chapter 4

WHY WE SHOULD CARE ABOUT THE PARASITE FAUNA OF CETACEANS: A PLEA FOR INTEGRATIVE STUDIES

F.J. Aznar, M. Fernández[†] and J.A Balbuena[*]
Marine Zoology Unit, Cavanilles Institute of Biodiversity and Evolutionary Biology and Fundación General de la UVEG, University of Valencia,
Valencia, Spain

ABSTRACT

Cetaceans harbor a unique and apparently little diversified fauna of parasites (including commensals). Unlike their hosts, none of these whales' dwellers currently benefit from any conservation measure, perhaps because they are unappealing to humans, often cause harm to their hosts, and many can easily go unnoticed due to of their small size. However, we can put forward a convincing point for promoting conservation of these creatures: parasites can provide a wealth of insight into the evolutionary and demographic history of whales, two aspects of prime importance for conservation. For instance, genetic information of whale-lice has recently been used to reconstruct with great accuracy the population histories of the highly endangered species of right whales.

---

* P.O. Box 22085, 46071 Valencia, Spain.
†Cavanilles Institute of Biodiversity and Evolutionary Biology, and Fundación General de la Universidad de Valencia

Likewise, our preliminary study of the dolphin-dweller barnacle *Xenobalanus globicipitis* suggests that Mediterranean populations are strongly impacted by striped-dolphin die-offs. If this were further confirmed, it would open the possibility of tracking the history of high dolphin mortality events by using molecular markers of the barnacles. Parasites have also the potential to reconstruct ecological scenarios of cetaceans. For instance, recent evidence shows that a small group of tetraphyllidean cestodes use cetaceans as intermediate hosts and large sharks as definitive hosts. These parasites might, therefore, be useful to investigate the origin of the trophic interactions between sharks and cetaceans. Convincing as the above may seem, we could additionally argue that the parasites of cetaceans, like any other organism, are masterpieces of evolution that deserve equal conservation 'rights' as their hosts, hence the current concept of co-extinction risk. Indeed some of them represent fascinating examples of evolutionary novelty (e.g., barnacles settling on living tissues or copepods apparently able to feed on the keratin of baleen plates). From this holistic perspective, how could we scientists help preserving both cetaceans and their parasites? Among other measures, parasites should be sampled whenever the opportunity arises (e.g., from stranded cetaceans). Also, sampling protocols and research studies should (i) consider parasites beyond their pathogenic impact; (ii) be conceived within collaborative networks, and (iii) be adapted to the resource levels and time available of the teams involved.

Do Parasites Matter?

On the 3rd of May, 2009, somewhere off the Mediterranean coast of Spain, a freighter struck and killed an adult 18m-long fin whale, *Balaenoptera physalus*. The animal was inadvertently dragged by the ship to the harbor of Valencia, where the Port Authority realized the unfortunate (but not too unusual) incident and called to our lab for biological sampling and (perhaps more importantly for them) directions on carcass disposal. Sampling was clearly constrained by shortage of time and the animal's huge size. So our on-site decision was to measure the animal, take samples of skin, blubber and muscle (useful for genetic and ecotoxicological studies), and collect as many ectoparasites as possible. In fact, we were as interested in the ectoparasites as in the whale itself, and this was not just a matter of "expertise bias" (we are primarily a parasitology lab), but there were strong additional reasons for focusing on parasites.

In this critical comment, we will briefly discuss why the parasites of cetaceans (for brevity, we will use the term "parasite" to refer to both metazoan true parasites and commensals) should matter, not only to parasitologists, but also to whale scientists, and why parasites should be included in sampling protocols and used more frequently to uncover aspects of their hosts' biology. In addition, we will provide some basic recommendations to conduct sampling of these organisms so that their scientific utility is maximized.

PARASITES AND VALUES

Public perceptions of species have a major influence on allocation of research effort as well as on formulation and implementation of conservation policies. These perceptions are embedded in instrumental, eudaemonic (aesthetic) and moral values that are reflected in the 'social construction' of species (Czech et al. 1998). Most mammals, including cetaceans, are subjects of much research and receive policy benefits because they have a positive social construction derived from strongly positive values. They clearly fall within the category of sentient beings thereby entitled to certain legal and moral rights (see, e.g., EU 1997). Conversely, parasites are negatively constructed by society: they cause harm to their sentient hosts, and are usually perceived as small and creepy. Thus, parasites attract much less research funds and fewer students, utterly lack political advocates, and receive no direct conservation benefits. In fact, only one metazoan parasite (a species of louse) is listed on the IUCN Red List of Threatened Species (Whiteman and Parker 2005). Although one could argue that parasites benefit indirectly from the conservation measures applied to their hosts, this not necessarily guards parasites against extinction (Whiteman and Parker 2005). On the other hand, endangered parasites have intentionally been killed in efforts to recover populations of their endangered hosts (Gompper and Williams 1998, Koh et al. 2004).

Using the provocative slogan "Equal Rights for Parasites", Windsor (1995) attempted to call attention to the neglected moral and instrumental values of parasites. About their moral values, Windsor (1995) claimed: "if we truly appreciate biological diversity, we must advocate that all species are precious, even parasites". This statement tracks Wilson's (1992) view that any extant species is a masterpiece of evolution, and also dispels Orwell's famous cynical motto in *Animal Farm*: "all animals are equal, but some animals are more equal than others". In summary, Windsor (1995)'s view is a form of moral biospheric value (see Stern 2000).

With regard to instrumental values, Windsor (1995) argued that parasites are among the most diverse of all organisms and are integral parts of ecosystem functioning. In fact, this field of research has blossomed over the last decade and has demonstrated that parasitism is not simply a pathogenic relationship requiring treatment, but a process of paramount importance in generating diversity, stability and resilience of ecosystems (Combes 1996, Horwitz and Wilcox 2005, Hudson et al. 2006, and references therein). In Hudson et al. (2006)'s words: "a healthy ecosystem is an infected ecosystem".

Another positive instrumental dimension of parasites concerns their use as natural tags to infer biological features of their hosts, such as social behavior, diet, migration, distribution, dispersal, stock identity, phylogeography, phylogeny, and disease. This field of study has a relatively long tradition; Williams, MacKenzie et al. (1992) listed over 190 papers on the subject only in the 1970s and 1980s. The suitability of parasites as tags ultimately relies on the intimate dependency that parasites establish with their hosts and the fact that transmission is linked to local features of both habitats and host populations. Classical studies of parasites as tags are based on presence/absence data or counts of suitable parasite taxa and, therefore, they represent a cost-effective method (Mackenzie 2002, and references therein). Often, these parasite studies work as subsidiary confirmatory tool of other methods (e.g., molecular markers), but this is not always the case. For instance, differences in parasite loads among localities have indicated temporal isolation of fish stocks where molecular markers failed to find population subdivisions, and this information may be crucial for proper management of fish stocks (Lester, 1990).

With the advent of molecular techniques, the use of parasites as tags is growing steadily. Parasites share a common history with their hosts, and their genotypes have often been shown to provide a better resolution of this common history than those of the host itself (see Nieberding and Olivieri, 2007 for a review). To mention some recent examples, the genotype of parasites has proven to be more useful than that of their hosts to reveal the phylogeographic history of rodents (Nieberding et al. 2004), to understand dispersal patterns of sea gulls (McCoy et al. 2005), or to assign salmons to their population of origin (Criscione et al. 2006). In fact, Whiteman and Parker (2005) brought forward these advantages as a pragmatic rationale for parasite conservation. In any event, these studies should obviously be viewed as a complement, not as a substitute, of research on the hosts, but the point to be stressed is that considering the whole association is by far scientifically more productive than just focusing on one of the partners.

THE VALUES OF PARASITES OF CETACEANS

Cetaceans are peculiar among all mammals (except sirenians) in that they have colonized marine waters severing their ties with land. The transition from land to sea likely posed a barrier for terrestrial parasites to track the cetacean ancestors. Indeed, most, if not all, metazoan parasite lineages in extant cetaceans have likely been acquired from other marine hosts (Aznar et al. 2001, Raga et al. 2009). As a result, the present-day parasite fauna of cetaceans seem to be relatively poor and highly specific (Aznar et al. 2001, Raga et al. 2009). Note, however, that diversity estimates are strongly affected by two factors (Raga et al. 2009). First, there are obvious difficulties for an appropriate sampling of species. For instance, beaked whales (Ziphiidae) represent one the most speciose families of cetaceans but they live in the high seas and have seldom been surveyed for parasites. Second, many parasite species that are currently considered as cosmopolitan or widespread may actually represent complexes of sibling or pseudosibling species.

Moreover, a number of unique and fascinating morphological and ecological peculiarities have occurred during the co-adaptation process of parasites to cetaceans. Consider, for example, the barnacles of the family Coronulidae, which have managed to settle on whales and dolphins by anchoring within the skin; none of the glues typically used by any other barnacle would ever work on a shedding substrate as the skin (Seilacher 2005). *Balaenophilus unisetus*, a harpacticoid copepod from whales, is apparently able to feed on the keratin of baleens, a exceptional food resource among crustaceans (Badillo et al. 2007). *Placentonema gigantissima* dwells in the placenta of sperm whales, *Physeter macrocephalus*, being the largest nematode known (females reach 9 m in length by 2.5 cm in width) (Dailey 1985): how and why this species became gigantic (see McClain and Boyer 2009)? How could it fit within the observed evolutionary patterns of life history traits of nematodes (Morand and Sorci, 1998)? Evidence also suggests that cetaceans act as intermediate hosts for tetraphyllidean cestodes that become adults in large lamnid sharks worldwide (Aznar et al. 2007). This represents a rather strange case of a parasite's life cycle that historically ended up involving mammals as intermediate hosts and fish as definitive hosts.

The above examples illustrate two points. First, if we adhere to the argument that any extant species is intrinsically valuable, both cetaceans and their parasites should be viewed as examples of the wealth jeopardized by current loss of biodiversity. Second, if we agree that uniqueness is a major criterion for setting conservation priorities of taxa (see e.g., Pyenson 2009), we should remember that not only cetaceans, but also their parasites, are sources of evolutionary novelty. Following this argument, one must wonder why our justified worries about the Yangtze River dolphin, *Lipotes vexillifer*, (now likely to be extinct) (Turvey et al. 2007, Pyenson 2009), and the vaquita, *Phocoena sinus*, (critically endangered) (Jaramillo-Legorreta et al. 2007) do not encompass also their parasites: could somebody say whether there are parasites specific to these hosts that have become co-extinct or are critically co-endangered?

Parasites have also been used as biological tags of cetaceans (Balbuena et al. 1995). Using qualitative or quantitative data on single or multiple parasite taxa, it has been possible to obtain evidence about host movements (or lack of them) (Abril et al. 1986, Aznar et al. 1995), stock structure (Dailey and Vogelbein 1991, Walker 2001), social behavior and structure (Best 1969, Balbuena and Raga 1991, 1993), and disease (Aznar et al. 1994, 2005). Admittedly, these studies were generally confirmatory, but they also have the potential for original insight. For instance, mass infections of the whale-lice *Isocyamus delphini* occur only on unhealed wounds on the bulls of long-finned pilot whale, *Globicephala melas*, in NE Atlantic (Balbuena and Raga, 1991), and perhaps these infections could be used to identify dominant males in sighting surveys (Balbuena et al. 1995). Likewise, epizoic crustaceans were more frequently observed in striped dolphins, *Stenella coeruleoalba*, affected by an immunosupressive viral infection (Aznar et al. 2005). This opens the possibility for these crustaceans to be used to assist health monitoring of wild cetacean populations (Aznar et al. 2005).

As far as we are aware, molecular methods have been used only once to infer aspects of cetacean biology based on the genotype of their parasites. Kaliszewska et al. (2005) used the mitochondrial *COI* gen from 3 whale-louse species obtained from stranded hosts to investigate the population history of right whales, *Eubalaena glacialis*, *E. japonica* and *E. australis*. Major findings were that (1) all 3 whale-louse species from the N Atlantic and Southern Ocean formed reciprocally monophyletic clades, which implies that whales from each ocean have long been isolated; (2) there were consistent divergence times of sibling species pairs in the 3 whale-louse species, which falls within the range (3-12 Ma) estimated for North Atlantic and Southern right whales; and (3) there was no detectable differentiation in any whale-louse species on right whales from the major Southern Ocean breeding aggregations, which suggests that there are frequent contacts among whales from all aggregations. In summary, whale-lice represent independent 'replicate evolutionary experiments' that shed much light on the evolutionary history of their hosts (Kaliszewska et al. 2005).

The above study highlights the great potential of molecular markers applied to parasites for population and community studies of cetaceans at multiple temporal and spatial scales. To mention a couple of examples drawn from our own research, there is preliminary evidence that the mass mortalities suffered by Mediterranean striped dolphins, *Stenella coeruleoalba*, (Raga et al. 2008) could also have affected the population of the epizoic barnacle *Xenobalanus globicipitis* (unpub. data). The question that arises is whether bottleneck test methods could be used to look for a signature of recent population size changes in genetic variation of the barnacle, perhaps coupled with, or not detectable in, those of its host. Moreover, the relationship between tetraphyllidean cestodes and cetaceans (see above) suggests the existence of long-term, predictable predator-prey interactions between large sharks and cetaceans (Aznar et al. 2007). When did these interactions begin? The use of appropriate molecular markers applied to these tapeworms could shed light on this question.

A Plea for Integrative Studies

We hope that the above discussion is compelling enough and the reader is now convinced about the value of considering both cetaceans and their parasites in sampling protocols and scientific studies. Actually, detailed directions for sampling of parasites is currently included in the majority of protocols for study of stranded cetaceans worldwide (see, e.g., Kuiken and García Hartmman 1993, Geraci and Lounsbury 2005, Australian Government 2006, Pugliares et al. 2007, Rommel et al. 2007), but the focus is largely on parasites as pathogens. The present critical comment attempts to broaden the perception about the moral and (positive) instrumental values of parasites, which has further consequences in the way they should scientifically be viewed.

With regard to sampling protocols for stranded cetaceans, three formal criteria could help improving decisions about what, and how many, parasites to collect:

1. *Rarity of the cetacean species*. Regarding moral values, new parasite taxa are intrinsically valuable (a biospheric value) and should join the biodiversity inventory as soon as possible. Accordingly, researchers

should strive to collect as many parasites as possible in poorly surveyed cetacean species because (1) there is a higher probability of finding new parasite taxa in these hosts, and (2) taxonomic studies on these parasites would ideally require a large number of well-preserved specimens. Beaked whales are currently the first candidates to be considered as 'rare' hosts in this sense.

With regard to instrumental values, parasites from poorly-surveyed hosts may have the key to illuminating particular aspects of the hosts' biology. For instance, the inferences made by the study of whale-lice on right whales discussed above were limited by the shortage of samples from North Pacific right whales, *Eubalaena japonica* (see Kaliszewska et al. 2005). Clearly, the next stranded whale of this species will be very valuable!

2. *Resource levels of the team, including personnel and facilities*. There is a growing call for concerted efforts worldwide to coordinate sampling and investigation of cetaceans and other marine mammals (see, e.g., Gulland and Hall 2007). Concerning parasites, our general recommendation for whale scientists is simple. First, consult with microbiologists, parasitologists or other specialists before proceeding with large-scale sampling of these organisms. Second, when possible, freeze entire organs (particularly the gut and lungs) of stranded cetaceans, unless fresh parasite species are needed (see, e.g., Point 1), then contact specialists for advice and/or future collaboration. When freezing is not feasible, try to collect as many parasites as possible, following the guidelines provided in specialized manuals (e.g., Geraci and Lounsbury 2005). Preserve at least some specimens in 70% ethanol since formaline is unsuitable for molecular studies (see e.g., Williams, Pontén et al. 1999).

   Obviously, research resources are often limited, particularly in developing countries. Still, any stranded cetacean provides an excellent opportunity to get very useful parasite samples with the only aid of a knife, a forceps, some vials, and a modest amount of ethanol; these samples could be included in large-scale collaborative work (see, e.g., Kaliszewska et al. 2005).
3. *Time available*. In the real world, time is often the scarcest resource. Therefore, in host-centered studies, it is often not possible to focus on parasites in depth. In these cases, it may be hard deciding what species, and how many specimens, of parasites should be collected. Decisions will obviously depend on the specific study or collaborative

research one has in mind, and the circumstances of sampling (as in our fin-whale example above). However, general tips for a cost-effective sampling can be proposed. First, many parasite-based inferences useful for cetacean populations rely on molecular data and, therefore, there will be no need for complete sampling of parasite populations (note, however, that molecular methods are not the panacea—see above). Second, some parasites are generally better candidates to trace the genealogical history of their hosts' populations, i.e., parasites that are host-specific, exhibit simple life cycles with limited horizontal transmission, and have smaller effective population sizes and shorter generation times than their hosts (Nieberding and Olivieri 2007). Among parasites of cetaceans, some ectoparasites (and ectosymbionts) and lungworms (Pseudaliidae) would appear the first candidates for testing.

As a closing thought, let's return to the day the unfortunate fin whale arrived at the Spanish harbor. We eventually found just a single whale-louse on the skin of the 50-ton whale. We are sure, however, that this solitary traveler has much to tell about ecology and evolution in ways yet to be understood.

REFERENCES

Abril, E., Almor, P., Raga, J. A., and Duguy, R. (1986). Parasitisme par *Anisakis typica* (Diesing, 1860) chez le dauphin bleu et blanc (*Stenella coeruleoalba*) dans le Nord-Est Atlantique. *Bulletin de la Société Zoologique de France*, *111*, 131-133.

Australian Government (2006). Standardised protocols for the collection of biological samples from stranded cetaceans. 65 pp. (http://www.environment.gov.au/coasts/publications/cetacean-protocols/pubs/cetacean-protocols.pdf)

Aznar, F. J., Agustí, C., Littlewood, D. T. J., Raga, J. A., and Olson, P. D. (2007). Insight into the role of cetaceans in the life cycle of the tetraphyllideans (Platyhelminthes: Cestoda). *International Journal for Parasitology*, *37*, 243–255.

Aznar, F. J., Balbuena, J. A., and Raga, J. A. (1994). Are epizoites biological indicators of a western Mediterranean striped dolphin die-off? *Diseases of Aquatic Organisms*, *18*, 159-163.

Aznar, F. J., Balbuena, J. A., Fernández , M., and Raga, J. A. (2001). Living together: the parasites of marine mammals. In P. G. H. Evans, and J. A. Raga (Eds.), *Marine Mammals: Biology and Conservation* (pp. 385-423). New York: Kluwer Academic/Plenum Publishers.

Aznar, F. J., Perdiguero, D., Pérez del Olmo, A., Repullés, A., Agustí, C., and Raga, J. A. (2005). Changes in epizoic infections during cetacean die-offs: the mass mortality of Mediterranean striped dolphins, *Stenella coeruleoalba*, revisited. *Diseases of Aquatic Organisms*, *67,* 239-247.

Aznar, F. J., Raga, J. A., Corcuera, J., and Monzon, F. (1995). Helminths as biological tags for franciscana *Pontoporia blainvillei* (Cetacea, Pontoporiidae) in Argentinean and Uruguayan waters. *Mammalia*, *59*, 427-435.

Badillo, F. J., Puig, L., Montero, F. E., Raga, J. A. and Aznar, F. J. (2007). Diet of *Balaenophilus* spp. (Copepoda: Harpacticoida): feeding on keratin at sea? *Marine Biology*, *151*, 751-758.

Balbuena, J. A., Aznar, F. J., Fernández, M., and Raga, J. A. (1995). The use of parasites as indicators of social structure and stock identity of marine mammals. In A. S. Blix, L. Walløe, and Ø. Ulltang (Eds.), *Whales, seals, fish and man* (pp. 133-139). Amsterdam: Elsevier Science.

Balbuena, J. A., and Raga, J. A. (1991). Ecology and host relationships of the whale-louse *Isocyamus delphini* (Amphipoda: Cyamidae) parasitizing long-finned pilot whales (*Globicephala melas*) off the Faroe Islands (N.E. Atlantic). *Canadian Journal of Zoology*, *69*, 141- 145.

Balbuena, J. A., and Raga, J. A. (1993). Intestinal helminth communities of the long-finned pilot whale (*Globicephala melas*) off the Faroe Islands. *Parasitology*, *106*, 327-333.

Best, P. B. (1969). The sperm whale (*Physeter catodon*) off the west coast of South Africa 3. Reproduction in the male. *Investigational report of the Division of Sea Fisheries of South Africa*, *72*, 1-20.

Combes, C. (1996). Parasites, biodiversity and ecosystem stability. *Biodiversity and Conservation*, *5*, 953-962.

Criscione, C. D., Cooper, B., and Blouin, M. S. (2006). Parasite genotypes identify source populations of migratory fish more accurately than fish genotypes. *Ecology*, *87*, 823-828.

Czech, B., Krausman, P. R., and Borkhataria, R. (1998). Social construction, political power, and the allocation of benefits to endangered species. *Conservation Biology*, *12*, 1103-1112.

Dailey, M. D. (1985). Diseases of Mammalia: Cetacea. In O. Kinne (Ed.), *Diseases of Marine Animals: Introduction, Reptilia, Aves, Mammalia,*

Volume IV, Part 2 (pp. 805–847). Hamburg: Biologische Anstalt Helgoland.

Dailey, M. D., and Vogelbein, W. K. (1991). Parasite fauna of three species of Antarctic whales with reference to their use as potential stock indicators. *Fishery Bulletin*, *89,* 355-365.

EU (1997). Protocol on protection and welfare of animals. Treaty of Amsterdam amending the Treaty on European Union, the Treaties establishing the European Communities and related acts. *Official Journal C 340.*

Geraci, J. R., and Lounsbury, V. L. (2005). *Marine mammals ashore: a field guide for strandings* (2nd edition). Baltimore, MD: National Aquarium in Baltimore.

Gompper, M. E., and Williams, E. S. (1998). Parasite conservation and the black-footed ferret recovery program. *Conservation Biology*, *12*, 730-732.

Gulland, F. M. D., and Hall, A. J. (2007). Is marine mammal health deteriorating? Trends in the global reporting of marine mammal disease. *EcoHealth*, *4*, 135-150.

Horwitz, P., and Wilcox, B. A. (2005). Parasites, ecosystems and sustainability: an ecological and complex systems perspective. *International Journal for Parasitology*, *35*, 725-732.

Hudson, P. J., Dobson, A. P., and Lafferty, K. D. (2006). Is a healthy ecosystem one that is rich in parasites? *Trends in Ecology and Evolution*, *21*, 381-385.

Jaramillo-Legorreta, A., Rojas-Bracho, L., Brownell, R. L. Jr , Read, A. J., Reeves, R. R., Ralls, K., and Taylor, B. L. (2007). Saving the vaquita: Immediate action, not more data. *Conservation Biology, 21*, 1653-1655.

Kaliszewska, Z. A., Seger, J., Rowntree, V. J., Barco, S. G., Benegas, R., Best, P. B., Brown, M. W., Brownell Jr., R. L., Carribero, A., Harcourt, R., Knowlton, A. R., Marshall-Tilas, K., Patenaude, N. J., Rivarola, M., Schaeff, C. M., Sironi, M., Smith, W. A., and Yamada, T. K. (2005). Population histories of right whales (Cetacea: *Eubalaena*) inferred from mitochondrial sequence diversities and divergences of their whale lice (Amphipoda: *Cyamus*). *Molecular Ecology*, *14*, 3439-3456.

Koh, L. P., Dunn, R. R., Sodhi, N. S., Colwell, R. K., Proctor, H. C., and Smith, V. S. (2004). Species coextinctions and the biodiversity crisis. *Science*, *305*, 1632-1634.

Kuiken, M. and García Hartmann, M. (1993). Proceedings of the first ECS workshop on cetacean pathology: Dissection techniques and tissue sampling. *ECS Special Newsletter*, *17*, 26-39.

Lester, R. J. G. (1990). Reappraisal of the use of parasites for fish stock identification. *Australian Journal of Marine and Freshwater Research*, *41*, 855–864.

Mackenzie, K. (2002). Parasites as biological tags in population studies of marine organisms: an update. *Parasitology*, *124*, S153-S163.

McClain, C. R., and Boyer, A. G. (2009). Biodiversity and body size are linked across metazoans. *Proceedings of the Royal Society B*, *276*, 2209-2215.

McCoy, K. D., Boulinier, T., and Tirard, C. (2005). Comparative host-parasite population structures: disentangling prospecting and dispersal in the black-legged kittiwake *Rissa tridactyla*. *Molecular Ecology*, *14*, 2825-2838.

Morand S., and Sorci, G. (1998). Determinants of life-history evolution in nematodes. *Parasitology Today*, *14*, 193-196.

Nieberding, C., Morand, S., Libois, R., and Michaux, J. (2004). A parasite reveals cryptic phylogeographic history of its host. *Proceedings of the Royal Society of London B*, *271*, 2559-2568.

Nieberding, C. and Olivieri, I. (2007). Parasites: proxies for host history and ecology? *Trends in Ecology and Evolution*, *22*, 156-165.

Pugliares, K. R., Bogomolni, A., Touhey, K. M., Herzig, S. M., Harry, C. T., and Moore, M. J. (2007). Marine mammal necropsy: An introductory guide for stranding responders and field biologists. *Technical Report WHOI-2007-06* (1-131). Woods Hole Oceanographic Institution. (https://darchive.mblwhoilibrary.org/bitstream/1912/1823/3/WHOI-2007-06.pdf)

Pyenson, N. D. (2009). Requiem for *Lipotes*: An evolutionary perspective on marine mammal extinction. *Marine Mammal Science*, *25,* 714-724.

Raga, J. A., Fernández, M., Balbuena, J. A., and Aznar, F. J. (2009). Parasites. In W. F. Perrin, B. Würsig, and J. G. M. Thewissen (Eds.), *Encyclopedia of Marine Mammals* (2nd edition, pp. 821-830). New York: Academic Press.

Raga, J. A., Banyard, A., Domingo, M., Corteyn, M., Van Bressem, M. F., Fernández, M., Aznar, F. J., and Barrett, T. (2008). Dolphin morbillivirus epizootic resurges in the Mediterranean. *Emerging Infectious Diseases*, *14*, 471-473.

Rommel, S., McLellan, W., Costidis, A., and Moore, M. (2007). Odontocete necropsy procedure. Chapter 2. In N. M. Young (Ed.), *Odontocete Salvage, Necropsy, Ear Extraction, and Imaging Protocols* (1-64). New techniques to diagnose possible noise-related strandings workshop. (http://www.nmfs.noaa.gov/pr/health/noise/docs/protocols.pdf)

Seilacher, A. (2005). Whale barnacles: exaptational access to a forbidden paradise. *Paleobiology*, *31*, 27-35.

Stern, P. C. (2000). Toward a coherent theory of environmentally significant behavior. *Journal of Social Issues*, *56*, 407-424.

Turvey, S. T., Pitman, R. L., Taylor, B. L., Barlow, J., Akamatsu, T., Barrett, L. A., Zhao, X., Reeves, R. R., Stewart, B. S., Wang, K., Wei, Z., Zhang, X., Pusser, L. T., Richlen, M., Brandon, J. R., and Wang, D. (2007). First human-caused extinction of a cetacean species? *Biology Letters*, *3*, 537-540.

Walker, W. A. (2001). Geographical variation of the parasite, *Phyllobothrium delphini* (Cestoda), in Dall's porpoise, *Phocoenoides dalli*, in the northern North Pacific, Bering Sea, and Sea of Okhotsk. *Marine Mammal Science*, *17*, 264-275.

Whiteman, N. K., and Parker, P. G. (2005). Using parasites to infer host population history: A new rationale for parasite conservation. *Animal Conservation*, *8*, 175-181.

Williams, H. H., MacKenzie, K., and McCarthy, A. M. (1992). Parasites as biological indicators of the population biology, migrations, diet, and phylogenetics of fish. *Reviews in Fish Biology and Fisheries*, *2*, 144–176.

Williams, C., Pontén, F., Moberg, C., Söderkvist, P., Uhlén, M., Pontén, J., Sitbon, G., and Lundeberg, J. (1999). A high frequency of sequence alterations is due to formalin fixation of archival specimens. *American Journal of Pathology*, *155*, 1467–1471.

Wilson, E. O. (1992). *The diversity of life*. Cambridge, Massachusetts: Harvard University Press.

Windsor, D. A. (1995). Equal right for parasites. *Conservation Biology*, *9*, 1-2.

In: Whales and Dolphins
Editor: Craig A. Murray, pp. 127-138
ISBN 978-1-61668-693-2

Chapter 5

MORE THAN JUST OLD BONES: THE FOSSIL RECORD INFORMS US ABOUT HOMOLOGY, AND CONVERGENCES OF ANATOMY, PHYSIOLOGY AND BEHAVIOR IN THE CETACEA OTHERWISE UNKNOWN FROM MODERN ANIMALS

Brian Lee Beatty*[1]• *and Alton C. Dooley, Jr.* [2]

1. Department of Anatomy,
New York College of Osteopathic Medicine,
Old Westbury, New York, USA
2. Virginia Museum of Natural History,
Martinsville, Virginia, USA

ABSTRACT

The biology of cetaceans is one of the most compelling because of the extreme adaptations whales and dolphins have had to evolve to manage a life in water. The fossil record of cetaceans is rich, and though much attention has been given to the origins of whales from terrestrial artiodactyls, it is important to realize that the biology, physiology, and

---

• email: bbeatty@nyit.edu.
∗ email: Alton.dooley@vmnh.virginia.gov.

behavior of modern cetaceans has not remained unchanged since this initial transition to being aquatic. Here I review some examples of how the fossil record of cetaceans informs us of how the evolution of anatomy, physiology and behavior has diverged and converged between and within the Odontoceti and Mysticeti in ways that would not be known if one were only to study their living representatives. Studies of paleopathologies associated with decompression syndrome inform us that odontocetes and mysticetes independently evolved specializations for repetitive deep diving. Cross sectional anatomy of ribs from modern and fossil mysticetes indicates that mysticetes started out with hyperdense skeleton and were probably benthic feeding, only to converge on the osteoporosis-like state found among most modern mysticetes and odontocetes. In the end, these studies of fossil cetaceans highlight the fact that many of our assumptions about homologies of anatomy, physiology, and behavior in modern cetaceans may be misled by only studying modern cetaceans, and that interpretations of modern animal biological data that rely on these sorts of assumptions should be reconsidered.

INTRODUCTION

What the average person knows about the Cetacea is that there are whales and dolphins and that, thanks to efforts to get the word out about recent advances in science, they are most closely related to "cows" (artiodactyls). Yet the rich fossil record of cetaceans is often overlooked, and even when publicized, it is primarily the earliest fossil record of cetaceans, the transition from terrestrial to aquatic among the Archaeoceti, that is communicated (Prothero, 2007). Most of that transition occurred between 50-35 million years ago, leaving a large part of the evolution of the Cetacea remaining. In this time there were many different groups and forms (Fordyce and Muizon, 2001), yet outside the small world of marine mammal paleontology, these fossil cetaceans rarely get attention. Why?

I suspect that the reason for the lack of attention to the cetaceans that evolved after the terrestrial/aquatic transition is that a general lack of understanding exists about the differences between and among the constituents of the two modern groups of cetaceans, the Odontoceti and Mysticeti. In most cases, the most parsimonious assumption is that those few features found between these two groups, or within subsets of these two groups, are due to having been derived only once in the common ancestor of those groups.

ODONTOCETI VS. MYSTICETI

The obvious difference between the two groups is that the Odontoceti have teeth and are typically smaller than the Mysticeti, which are often larger and have baleen and no teeth. These two groups also share some commonalities, including having many members that dive deep. Due to the limitations of studying cetaceans, especially the large mysticetes or rarely encountered ziphiids, much of what we cannot readily observe from their anatomy and behavior must be inferred, basing it on what we know of related cetaceans. The direct datasets about physiology and behavior for ziphiids (Johnson et al., 2004; Tyack et al., 2006; Zimmer and Tyack, 2007) and mysticetes (Acevedo-Gutierrez et al., 2002; Croll et al., 2001; Goldbogen et al., 2006; Lagerquist et al., 2000; Nowacek et al., 2001; Stockin et al., 2001) is getting better, but for cases in which the physiological parameter being studied is something we cannot readily study in a large group of cetaceans that require such an inference of physiology, we must be cautious about what assumptions we are making about the homology of physiological characteristics inferred from animals that demonstrate similar behaviors. For example, in a study of decompression syndrome in mysticetes, (Houser et al., 2001), physiological data from mysticetes was plugged into models of respiratory physiology that were formulated from experimental data from *Tursiops truncatus*, a delphinid (Ridgway and Howard, 1979). This is not a fault of the researcher per se, as there are many, many hurdles that make invasive studies of cetacean physiology, especially in vivo, almost impossible. In the case of Houser et al. (2001), the authors go to great lengths to explain their assumptions and investigate what possible errors might result from them. In this way, sometimes assumptions about the homology of physiological parameters across large phylogenetic divergences are the only practical way to get things accomplished.

But a recent study of osteological evidence of dysbaric osteonecrosis in modern and fossil Cetacea, which is considered a type of evidence of repetitive deep diving and survival of repeated bouts of decompression syndrome, seems to indicate that repetitive deep diving behaviors may have evolved independently in the Odontoceti and Mysticeti (Beatty and Rothschild, 2008). In this study, Beatty and Rothschild (2008) make a case from a broad study of pathology in modern and fossil cetaceans, that dysbaric osteonecrosis, which is a pathology that results from repetitive deep diving of an air-breathing vertebrate that lacks either some or all of the physiological, anatomical, and/or behavioral specializations needed to avoid/minimize decompression syndrome. The argument they make is that, like humans repetitively diving deep and suffering these effects, those animals that perform the behaviors without the specializations required to do so regularly and avoid decompression syndrome would suffer from dysbaric osteonecrosis and show evidence of it in their fossil records at a higher frequency than those adapted to avoid those problems. These few ancient cetaceans that do show evidence of dysbaric osteonecrosis (the earliest odontocetes and some of the later mysticetes) are argued (Beatty and Rothschild, 2008) to have been individuals of the species that began the behavior of repetitive deep diving without the benefits of all the adaptations that modern cetaceans derived since their time. In a way, this is a glimpse of natural selection in action, and from the fossil record of cetaceans, this selection for specializations for repetitive deep diving appears to have occurred independently in the Odontoceti and Mysticeti after their divergence (Figure 1).

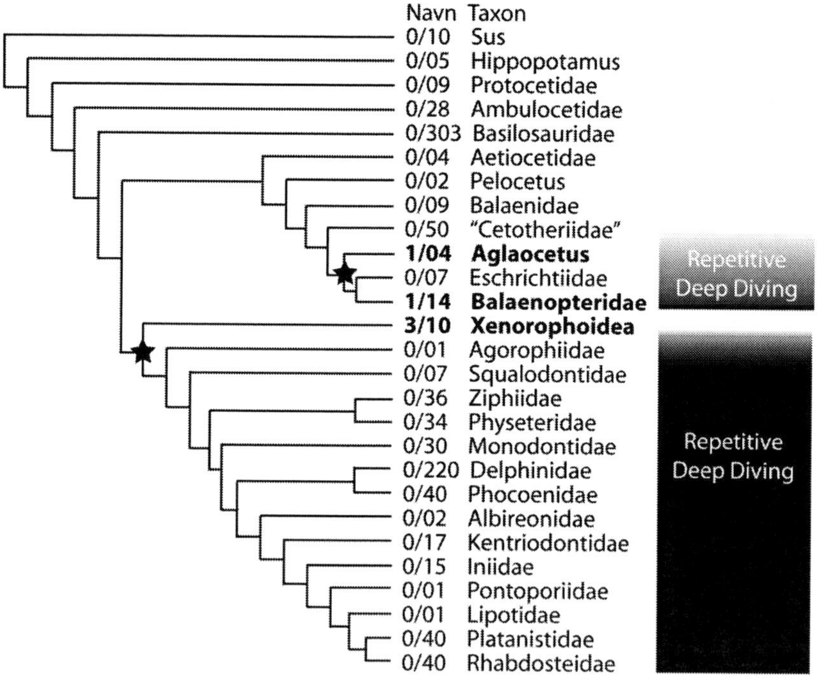

Figure 1. A phylogeny of the Cetacea with the incidence of avascular necrosis in modern and fossil specimens (modified from Beatty and Rothschild, 2008). 'Navn' is the frequency of individuals with avascular necrosis out of the fossil sample studied. The column at far right depicts how this is interpreted as indicative of the independent evolution of repetitive deep diving in the Odontoceti and Mysticeti.

Though this study cannot absolutely discern how these two groups initially accomplished avoidance of mortality from repetitive deep diving (whether it was accomplished initially by anatomical, physiological, or behavioral accommodation of the effects of deep diving), the simplest explanation of this data is still that these two groups independently came to repetitive deep diving. With this in mind, we must reexamine studies that are based on assumptions of this homology.

WITHIN THE MYSTICETI

Though the "pull of the recent" has a strict definition (Raup, 1979), it is commonly used as a general problem of the bias in collection and knowledge of the fossil record of geologically younger organisms (Pease, 1992). But the pull of the recent could also be interpreted as having an effect on our understanding of the diversity of the biology of organisms. For some time it has been well established that the most parsimonious inferences of the unknowable details of fossil species, such as soft tissues or behavior, are those based on optimization of such features on a phylogeny with modern members for which those features are known (Bryant and Russell, 1992). Therefore, the best inferences about modern mysticetes, if one were to only use modern representatives, is that fossil mysticetes shared most of the same soft tissue and behavioral characteristics that are found in both the modern mysticetes that are members of the same clade and those that are sister taxa to the clade containing the fossil taxa and other modern members.

If this is the case, then aspects of feeding and ecology of fossil mysticetes that are found within the modern clades of mysticetes should be shared as well. Modern mysticetes all lack teeth and bear baleen, specialized keratinized epidermal structures that allow them to filter feed. This sort of filter feeding is well known to differ among different groups, from the ram-feeding/skimming balaenids (Carroll et al., 1987; Laidre et al., 2007; Lambertsen et al., 2005; Rogachev et al., 2008; Sekiguchi et al., 1992; Watkins and Schevill, 1979; Werth, 2004; Werth, 2007) to the benthic feeding eschrichtiids (Darling et al., 1998; Dunham and Duffus, 2001; Nerini and Oliver, 1983; Oliver et al., 1984; Woodward and Winn, 2006), to the engulfment feeding balaenopterids (Arnold et al., 2005; Brodie, 2001; Goldbogen et al., 2006; Lambertsen, 1983; Lambertsen et al., 1995).

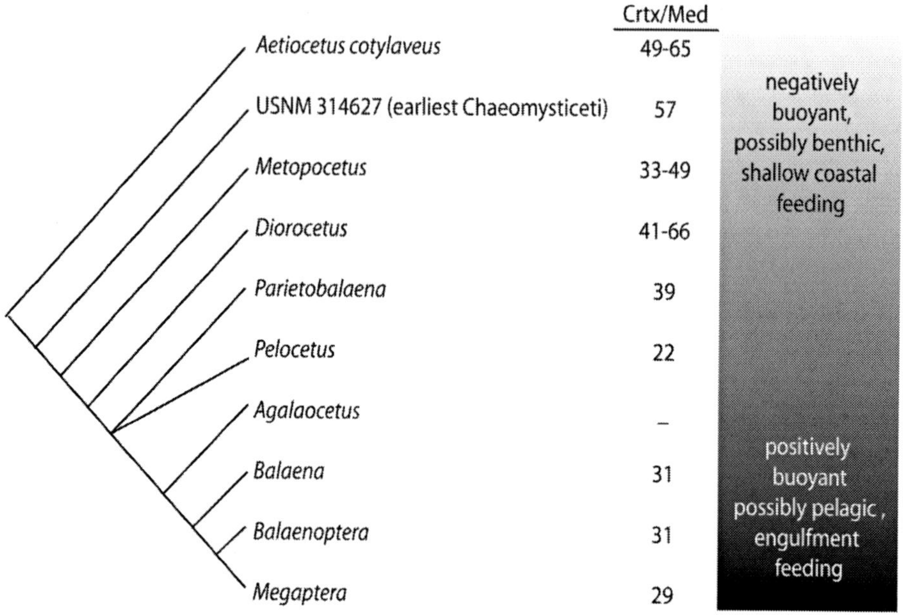

Figure 2. A phylogeny of the Mysticeti for which osteosclerosis data is known, derived from the phylogeny of Bouetel & de Muizon (2006). Crtx/Med is the ratio of cortical bone thickness to medullary bone thickness (see Beatty & Dooley, 2009 for more details on this). The column at far right depicts how this is interpreted as indicative of the transition of mysticetes from being negatively buoyant benthic feeders to positively buoyant, pelagic, and possibly engulfment feeders.

But the fossil record of mysticetes is rich, and the mysticetes alive today only represent a fraction of one of the more recent diversifications of the group in the Miocene (Fitzgerald, 2010; Fordyce and Muizon, 2001). Therefore, any inferences about the soft tissues and/or behavior of fossil mysticetes must ultimately be derived either from shared features between all mysticetes and their closest living relatives, odontocetes, or from interpretations of fossil mysticetes. Why is this important? Like outlined above, recognizing inferences based on such a deep divergence as that between the Mysticeti and Odontoceti may be fraught with convergence or not, and in many cases homologies may simply not exist. In inferring homology of traits it is a necessity to make assumptions of shared physiology/anatomy/behavior valid, and without modern representatives of the older clades of mysticetes those inferences become magnified.

Fossil representatives of the transition from tooth-bearing early whales to baleen-bearing mysticetes have been known since the late 1960s (Emlong, 1966), yet only recently have we begun to find out about the incredible diversity of these toothed mysticetes (Barnes et al., 1994; Deméré and Berta, 2008; Deméré et al., 2008; Emlong, 1966; Fitzgerald, 2006; Fitzgerald, 2010; Ichishima, 2005). Studies of how feeding ecology in this transition have reflected a gradation of changes from the use of teeth to the presence of baleen (Deméré et al., 2008), show that this transition occurred in a stepwise fashion early in mysticete evolution. Only very recently has it become clear that mysticetes started out with osteosclerotic rib cages that possibly made them benthic feeders (Beatty and Dooley, 2009).

But the phylogeny of Chaeomysticeti, the baleen-bearing mysticetes (Mitchell, 1989), is by no means stable, and if this stepwise evolution of baleen use and benthic feeding is optimized on these cladograms, the homology of feeding styles and any associated behavior, anatomy and/or physiology must be carefully examined. For example, if one plots the data of osteosclerosis (Beatty and Dooley, 2009) in one phylogeny (Bouetel and Muizon, 2006) one finds that the Cetotheriidae *sensu stricto*, as well as *Diorocetus*, maintain a high level of osteosclerosis like the earlier toothed mysticetes (*Aetiocetus cotylaveus* and an undescribed earliest Chaeomysticeti), indicating that they may all have been regular benthic feeders (Figure 2). This is particularly interesting in light of recent finds of a greater diversity of toothed mysticetes in the Southern Ocean (Fitzgerald, 2006; Fitzgerald, 2010), and that it implies that the primitive state for mysticetes, including the first baleen-bearing forms, was one of benthic feeding. Around the time of the divergence of *Pelocetus, Aglaeocetus*, and modern mysticete families (such as Balaenopteridae, Eschrichtiidae, and Balaenidae), this osteosclerosis seems to disappear. It is uncertain why this is the case, although it is interesting that the loss of osteosclerosis appears to occur at the same time as decompression syndrome shows up in *Aglaocetus* (Beatty and Rothschild, 2008), and when mysticetes evolve the specializations of their mandibles and skulls that allow them to perform the mechanically difficult and energetically expensive activity of engulfment feeding found in modern rorquals (Bouetel, 2005). This would seem to indicate that this group of crown mysticetes that has lost osteosclerosis is the same group that began repetitive deep diving and engulfment feeding. Although one might argue that it seems counterintuitive that osteosclerosis, an adaptation for negative buoyancy and benthic feeding, would be lost when mysticetes become deep divers, it makes perfect sense when compared to odontocetes. Although a complete study of bone histology in odontocetes has not been completed, from an early stage in evolution (Beatty, personal observation) odontocetes appear to have developed their unique form of the osteoporosis-like state so well known from modern taxa (Buffrénil et al., 1985; Buffrénil and Schoenvaert, 1988). That said, the data on dysbaric osteonecrosis indicates that the earliest odontocetes were engaging in repetitive deep diving (Beatty and Rothschild, 2008). Also, studies concerning the coevolution of odontocetes and diel migrating squid also appears to indicate an early evolution of deep diving for them (Lindberg and Pyenson, 2007). Thus, though all modern cetaceans appear to be mostly osteoporotic repetitive deep divers, the fossil record informs us that this may be convergent.

IMPLICATIONS

So what does this all mean? From sampling only modern cetaceans, the most parsimonious conclusion would be 1) the anatomical/physiological/ behavioral specializations present in both odontocetes and mysticetes are shared because they evolved once before the divergence of the Neoceti and 2) that the osteoporosis-like state of the bones of mysticetes is homologous with the similarly osteoporosis-like bones that are more well-known in odontocetes. But these are NOT the conclusions one would get from the added information from the fossil record of cetaceans. The fossil record informs us that in both of these cases, the osteoporosis-like state of bones and the specializations for repetitive deep diving of the Neoceti are convergences. Because of this, many of the assumptions we might work with when extrapolating data from one taxon to another may be false, and we need to investigate each of these assumptions of physiological, anatomical, and behavioral features much more carefully.

REFERENCES

Acevedo-Gutierrez, A., D. A. Croll, and B. R. Tershy. 2002. High feeding costs limit dive time in the largest whales. *The Journal of Experimental Biology*, 205:1747-1753.

Arnold, P. W., R. A. Birtles, S. Sobtzick, M. Matthews, and A. Dunstan. 2005. Gulping Behaviour in Rorqual Whales: Underwater Observations and

Functional Interpretation. *Memoirs of the Queensland Museum*, 51(2): 309-332.

Barnes, L. G., M. Kimura, H. Furusawa, and H. Sawamura. 1994. Classification and distribution of Oligocene Aetiocetidae (Mammalia; Cetacea; Mysticeti) from western North America and Japan. *The Island Arc*, 3(4):392-431.

Beatty, B. L., and A. C. Dooley. 2009. Injuries in a mysticete skeleton from the Miocene of Virginia, with a discussion of buoyancy and the primitive feeding mode in the Chaeomysticeti. *Jeffersoniana,* 20:1-29.

Beatty, B. L., and B. M. Rothschild. 2008. Decompression Syndrome and the Evolution of Deep Diving Physiology in the Cetacea. *Naturwissenschaften,* 95 (9):793-801.

Bouetel, V. 2005. Phylogenetic Implications of Skull Structure and Feeding Behavior in Balaenopterids (Cetacea, Mysticeti). *Journal of Mammalogy,* 86(1):139-146.

Bouetel, V., and C. d. Muizon. 2006. The anatomy and relationships of *Piscobalaena nana* (Cetacea, Mysticeti), a Cetotheriidae s.s. from the early Pliocene of Peru. *Geodiversitas*, 28(2):319-395.

Brodie, P. F. 2001. Feeding mechanics of rorquals (Balaenoptera sp.), p. 345-352. *In* J.-M. Mazin and V. d. Buffrénil (eds.), Secondary Adaptation of Tetrapods to Life in Water. Verlag Dr. Friedrich Pfeil, Munchen.

Bryant, H. N., and A. P. Russell. 1992. The Role of Phylogenetic Analysis in the Inference of Unpreserved Attributes of Extinct Taxa. *Philosophical Transactions of the Royal Society B: Biological Sciences*, 337(1282):405-418.

Buffrénil, V. d., A. S. Collet, and M. Pascal. 1985. Ontogenetic development of skeletal weight in a small delphinid, *Delphinus delphis* (Cetacea, Odontoceti). *Zoomorphology*, 105(5):336-344.

Buffrénil, V. D., And D. Schoenvaert. 1988. On How the Periosteal Bone of the Delphinid Humerus Becomes Cancellous: Ontogeny of a Histological Specialization. *Journal of Morphology*, 198:149-164.

Carroll, G. M., J. C. George, L. F. Lowry, and K. O. Coyle. 1987. Bowhead whale (*Balaena mysticetus*) feeding near Point Barrow, Alaska, during the 1985 spring migration. *Arctic*, 40(2):105-110.

Croll, D. A., A. Acevedo-Gutierrez, B. R. Tershy, and J. Urban-Ramrez. 2001. The diving behavior of blue and fin whales: is dive duration shorter than expected based on oxygen stores? Comparative Biochemistry and Physiology - Part A: *Molecular and Integrative Physiology*, 129:797-809.

Darling, J. D., K. E. Keogh, and T. E. Steeves. 1998. Gray whale (*Eschrichtius robustus*) habitat utilizations and prey species off Vancouver Island, B. C. *Marine Mammal Science*, 14(4):692-720.

Deméré, T. A., and A. Berta. 2008. Skull anatomy of the Oligocene toothed mysticete *Aetioceus weltoni* (Mammalia; Cetacea): implications for mysticete evolution and functional anatomy. *Zoological Journal of the Linnean Society,* 154(2):308-352.

Deméré, T. A., M. R. McGowen, A. Berta, and J. Gatesy. 2008. Morphological and Molecular Evidence for a Stepwise Evolutionary Transition from Teeth to Baleen in Mysticete Whales. *Systematic Biology,* 57(1):15 - 37.

Dunham, J. S., and D. A. Duffus. 2001. Foraging patterns of gray whales in central Calyoquot Sound, British Columbia, Canada. *Marine Ecology Progress Series*, 223:299-310.

Emlong, D. 1966. A new archaic cetacean from the Oligocene of Northwest Oregon. Bulletin of the Museum of Natural History, University of Oregon 3:1-51.

Fitzgerald, E. M. G. 2006. A bizarre new toothed mysticete (Cetacea) form Australia and the early evolution of baleen whales. *Proceedings of the Royal Society* B, 273(1604):2955-2963.

Fitzgerald, E. M. G. 2010. The morphology and systematics of *Mammalodon colliveri* (Cetacea: Mysticeti), a toothed mysticete from the Oligocene of Australia. *Zoological Journal of the Linnean Society* 158:367-476.

Fordyce, R. E., and C. d. Muizon. 2001. Evolutionary history of cetaceans: a review, p. 169-233. *In* J.-M. Mazin and V. d. Buffrénil (eds.), Secondary Adaptation of Tetrapods to Life in Water. Verlag Dr. Friedrich Pfeil, Munchen.

Goldbogen, J. A., J. Calambokidis, R. E. Shadwick, E. M. Oleson, M. A. McDonald, and J. A. Hildebrand. 2006. Kinematics of foraging dives and lunge-feeding in fin whales. *The Journal of Experimental Biology*, 209(7): 1231-1244.

Houser, D. S., R. Howard, and S. H. Ridgway. 2001. Can Diving-induced Tissue Nitrogen Supersaturation Increase the Chance of Acoustically Driven Bubble Growth in Marine Mammals? *Journal of Theoretical Biology*, 213:183-195.

Ichishima, H. 2005. Notes on the phyletic relationships of the Aetiocetidae and on the feeding ecology of toothed mysticetes. *Bulletin of Ashoro Museum of Paleontology,* 3:111-117.

Johnson, M., P. T. Madsen, W. M. X. Zimmer, N. A. de Soto, and P. L. Tyack. 2004. Beaked whales echolocate on prey. *Proceedings of the Royal Society B: Biological Sciences,* 271(0):S383-S386.

Lagerquist, B. A., K. M. Stafford, and B. R. Mate. 2000. Dive characteristics of satellite-monitored blue whales (*Balaenoptera musculus*) off the central California coast. *Marine Mammal Science*, 16:375-391.

Laidre, K. L., M. P. Heide-Jørgensen, and T. G. Nielsen. 2007. Role of the bowhead whale as a predator in West Greenland. *Marine Ecology Progress Series*, 346:285-297.

Lambertsen, R. H. 1983. Internal Mechanism of Rorqual Feeding. *Journal of Mammalogy*, 64(1):76-88.

Lambertsen, R. H., K. J. Rasmussen, W. C. Lancaster, and R. J. Hintz. 2005. Functional morphology of the mouth of the bowhead whale and its implications for conservation. *Journal of Mammalogy*, 86(2):342-352.

Lambertsen, R. H., N. Ulrich, and J. Straley. 1995. Frontomandibular Stay of Balaenopteridae: A Mechanism from Momentum Recapture During Feeding. *Journal of Mammalogy*, 76(3):877-899.

Lindberg, D. R., and N. D. Pyenson. 2007. Things that go bump in the night: evolutionary interactions between cephalopods and cetaceans in the tertiary. *LETHAIA,* 40(4):335-343.

Mitchell, E. D. 1989. A new cetacean from the Late Eocene La Meseta Formation, Seymour Island, Antarctic Peninsula. *Canadian Journal of Fisheries and Aquatic Sciences*, 46(10):2219-2235.

Nerini, M. K., and J. S. Oliver. 1983. Gray whales and the structure of the Bering Sea benthos. *Oecologia*, 59(2):224-225.

Nowacek, D. P., M. P. Johnson, P. L. Tyack, K. A. Shorter, W. A. McLellan, and D. A. Pabst. 2001. Buoyant balaenids: the ups and downs of buoyancy in right whales. *Proceedings of the Royal Society B: Biological Sciences*, 268(1478):1811-1816.

Oliver, J. S., P. N. Slattery, M. A. Silberstein, and E. F. O'Connor. 1984. Gray whale feeding on dense ampeliscid amphipod communities near Bamfield British Columbia. *Canadian Journal of Zoology*, 62:41-49.

Pease, C. M. 1992. On the Declining Extinction and Origination Rates of Fossil Taxa. *Paleobiology,* 18(1):89-92.

Prothero, D. R. 2007. Evolution - What the fossils say and why it matters. Columbia University Press, New York, 381 p.

Raup, D. M. 1979. Biases in the fossil record of species and genera. *Bulletin of the Carnegie Museum of Natural History*, 13:1-85.

Ridgway, S. H., and R. Howard. 1979. Dolphin lung collapse and intramuscular circulation during free diving: evidence from nitrogen washout. *Science,* 206:1651-1654.

Rogachev, K. A., E. C. Carmack, and M. G. G. Foreman. 2008. Bowhead whales feed on plankton concentrated by estuarine and tidal currents in Academy Bay, Sea of Okhotsk. *Continental Shelf Research*, 28(14):1811-1826.

Sekiguchi, K., P. B. Best, and B. Z. Kaczmaruk. 1992. New information on the feeding habits and baleen morphology of the pymy right whales *Caperea marginata*. *Marine Mammal Science*, 8(3):288-293.

Stockin, K. A., R. S. Fairbairns, E. C. M. Parsons, and D. W. Sims. 2001. Effects of diel and seasonal cycles on the dive duration of the minke whale (Balaenoptera acutorostrata). *Journal of the Marine Biology Association of the United Kingdom*, 81:189-190.

Tyack, P. L., M. Johnson, N. A. Soto, A. Sturlese, and P. T. Madsen. 2006. Extreme diving of beaked whales. *The Journal of Experimental Biology*, 209(21):4238-4253.

Watkins, W. A., and W. E. Schevill. 1979. Aerial Observation of Feeding Behavior in Four Baleen Whales: *Eubalaena glacialis, Balaenoptera borealis, Megaptera novaeangliae,* and *Balaenoptera physalus*. *Journal of Mammalogy*, 60(1):155-163

Werth, A. J. 2004. Models of hydrodynamic flow in the bowhard whale filter feeding apparatus. *The Journal of Experimental Biology*, 207:3569-3580.

Werth, A. J. 2007. Adaptations of the cetacean hyolingual apparatus for aquatic feeding and thermoregulation. The Anatomical Record: Advances in Integrative Anatomy and Evolutionary Biology, 290(6):546-568.

Woodward, B. L., and J. P. Winn. 2006. Apparent lateralized behavior in gray whales feeding off the central British Columbia coast. *Marine Mammal Science*, 22(1):64-73.

Zimmer, W. M. X., and P. L. Tyack. 2007. Repetitive Shallow Dives Pose Decompression Risk in Deep-Diving Beaked Whales. *Marine Mammal Science*, 23(4):888-925.

In: Whales and Dolphins
Editor: Craig A. Murray, pp. 139-150
ISBN 978-1-61668-693-2

Chapter 6

DIFFERENCES BETWEEN SOLITARY AND GROUP TIME BUDGET IN THE ESTUARINE DOLPHIN (*SOTALIA GUIANENSIS*) AT NORTHEASTERN – BRAZIL

***Carolina Herédia Tosi*[1•] *and Renata Gonçalves Ferreira*[2*]**
1. Pós Graduação em Psicobiologia,
Departamento de Fisiologia/Universidade Federal do Rio Grande do Norte – Natal/RN
2. Pós Graduação em Psicobiologia,
Departamento de Fisiologia/Universidade Federal do Rio Grande do Norte – Natal/RN

ABSTRACT

Advantages of social life relates to protection from predators and foraging efficiency. Disadvantages relate to the need to synchronise one's own activity with that of others. The analysis of which activity is facilitated or detracted in social groups offers an indication of the benefits and costs of solitary or social life, and the pressures individuals suffer in each situation. In this paper we compare the activity budget of adult estuarine dolphins (*Sotalia guianensis*) in two situations, solitary and in a

• E-mail: tosicarol@yahoo.com.br.
* E-mail: rgf27br@yahoo.com.br.

group. Analyses of 197 hours of behavioral sampling (Predominant Activity Sampling) at Pipa Bay located at south Rio Grande do Norte – Brazil support suggestions that protection from predators is facilitated in groups (animals Rest and Travel more in groups), but are obscure regarding advantages of groups for efficiency in foraging. The pressures for social life in estuarine dolphins are discussed in light of the small size of groups observed throughout its area of occurrence.

Keywords: *Sotalia guianensis, time budget, social life*

INTRODUCTION

Time or activity budget refers to the quantification of time employed by each individual in each activity needed for its survival or reproduction (Dunbar, 1988). Four basic categories are defined: Forage, Travel, Rest and Socialize, although more detailed activity budget may be used depending on research aims and hypothesis. Each individual has a theoretical optimum activity budget that when accomplished yields the maximum survival and reproduction with minimum energetic expenditure (Krebs and Davies, 1996).

Diverse factors hinder individuals from reaching their optimum activity budget, from ecological variables to social influences. Life in groups presents costs and benefits for the group members. Benefits are normally associated with protection from predators (Van schaick, 1983) and communal defense of feeding sites (Seyfarth, 1977), although other factors may also be important such as cooperative hunting, infant care and thermoregulation (Lee, 1994). Social life imposes costs either because similar organisms searching for similar resources are potential competitors or because each individual in a group has a different optimal foraging strategy. Life in groups, therefore, creates a constant need to compromise between costs and benefits, and it has been suggested that capacity for behavioural maneuvering in social life leads to increased cognition (Byrne and Whiten, 1988; Whiten and Byrne, 1997).

Odontocetes (Cetacea) are large brained aquatic mammals and their social life is related to protection from predators and, secondly, to foraging efficiency (Connor, 2007). Odontocetes exhibit diverse types of social organisation, from small groups, to matrilineal and fission-fusion societies (Mann et al., 2000). However, to our knowledge, no correlation between encephalisation index and size of groups was performed to odontocetes as it has been from primates (Dunbar, 1992).

The estuarine or grey dolphin (boto cinza), Sotalia guianensis (Van Bénédén, 1864), is the commonest dolphin in Brazilian shallow waters. It is a small odontocete (mean size of 1.8 meters) that forms small groups from 3 to 40 animals throughout its distribution (Geise et al., 1999; Lodi, 2003; Daura-Jorge et al., 2005; Azevedo et al., 2005; Queiroz, 2006). Larger groups are found in more protected bay than in coastal areas, probably due to the great availability of food in areas nearer river mouths than in open coastal waters (Queiroz et al., sub). However, large aggregations of animals (up 450 animals) were described (Lodi and Hetzel, 1998).

In this paper we present an analysis comparing the activity budget exhibited by adult *estuarine dolphins* (*Sotalia guianensis*) when solitary and when in adult groups during their daylight activity period. We assume that an analysis of which activity is facilitated or detracted in social groups offers an indication of the advantages and costs of solitary or social life, and the pressures individuals suffer in each situation.

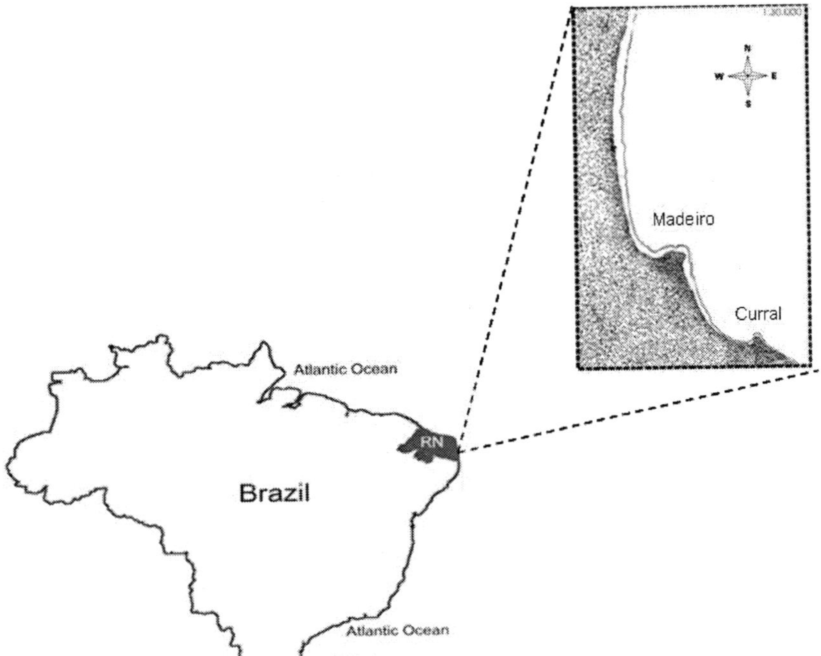

Figure 1. Localization of study area: Pipa Bay (Curral and Madeiro beach).

Materials and Methods

Study Areas

The study was conducted on two contiguous areas (Madeiro beach and Curral beach) that together compose the Pipa Bay area (6°13'21.16"S 35°04'15.91"W), located at south Rio Grande do Norte coast (figure 1). Pipa Bay presents a gradually shelving sea-bed, alternately rocky and sandy, with an average depth of 6m. The bay is encircled by cliffs approximately 30m in height, having sandy beaches which are heavily strewn in places with beachrocks, forming bays which are sheltered from the prevailing trade wind and ocean currents. Curral area has the remnants of an antique artisanal fishery area, and Madeiro area is frequented by tourists and surfing practitioners.

Data Collection

Direct observations were conducted from a land base located at Madeiro and Curral areas, with the help of binoculars (Bushnell 10x50), a digital clock, and specific register protocol. Daily observations lasted for 6 hours, during the period of February to December 2006. Accompaniment of dolphins occurred only during dry periods, with no data collections conducted on June, July and August, period of heavy rains. According to previous studies (Tosi and Ferreira, 2008; Guilherme-Silveira and Silva, 2007) dolphins forage less and locomote more after and before the rains.

Methodology of data registering was Predominant Activity Sampling (Mann and Smuts, 1999; Tosi and Ferreira, 2008) at two minutes intervals of each animal observed on the bay, regardless of the distance between them. At each registers it was noted: i) the behavioral state of each dolphin; ii) the number of dolphins within the bay.

Behavioral categories used in this work were adapted from many authors, and classified as: 1-Travel: swimming characterized by a uniform directional movement resulting on the change of animals' position from one area to another. It can occurs in different speeds; 2-Social Behavior: high level activity of the dolphins, where they remain longer on the surface and indulged in intense physical contact (cf. Balance, 1992; Mann and Smuts, 1999; Lusseau, 2003); 3-Rest: characterized by a pronounced diminution in one level of activity and very slow rate of swimming, without defined direction (cf. Würsig and Würsig 1980; Gnone et al., 2001; Constantine et al., 2004; Lusseau, 2003); 4-Forage: Involved variable directional movement, periodically interrupted by events of persecution and lunges towards a prey. This may or may not result in a successful catch of prey (cf. Norris and Dohl, 1980).

Data Analyses

The activity budget was calculated for the registers when only one adult within the bay (Solitary Activity Budget) and when more than one adult but no immature was present within the bay (Group Activity Budget). A group was defined as all individuals presented within the bay, regardless on the distance between them. We conducted pairwise analyses comparing solitary and social situation regardless of group size.

Data were analyzed using SPSS 13.0. When data were normally distributed T paired test was used, otherwise, Wilcoxon test for related samples was used. Significance value was set to 5% ($p \leq 0.05$) (Zar, 1996). Data were divided in two periods: before (February to May) and after the rains (September to December).

RESULTS

A total of 76 days of field effort and 70 days of effective data collection were accomplished, resulting in a total of 414.9 h of field effort and 197.12 h (47.5% success) of data collection. From these, 222.2 h of field effort or 88.66 h of effective data collection (39.9% success) occurred during the period 1 (before the rains), and 192.7 h of field effort and 110. 46 h of effective data collection (57.3% success) on period 2 (after the rains).

Time Budget

Period 1 – Travel occurs more often by groups of adults than by solitary dolphins (solitary: 38%; group: 48%), Rest is also more frequently exhibited by groups than by individuals alone (solitary: 2%; group: 5%). Conversely, Forage is more frequently exhibited when animals are alone within the bay than when in groups (solitary: 60%; groups: 41%). All differences were statistically significant: Travel ($t= -2.542$; $n= 31$ e $p= 0.016$), Rest ($z= -2.580$;

n= 31 e p= 0.010), Forage (t= 3.940; n= 31 e p= 0.000). Social behavior represents 5% of group activity budget (figure 2a).

Period 2 – After the rains the pattern of more solitary Forage (solitary: 79%; group: 67%) and more group Travel (solitary: 17%; group: 22%) and Rest (solitary: 3%; group: 5%) was maintained. However, only Forage presented a statistically significant difference: Forage: t= -3.594; n= 37 e p= 0.001; Travel: t= -1.397; n= 37 e p = 0.171; Rest: z= 1.402; n= 37 e p = 0.161). Social behavior represents 6% of group activity budget (figure 2b).

Comparison between Periods

On the period after the heavy rains individuals forage more, both alone (t= 3.261; n= 31 e p= 0.003) and in groups (t= -5.512; n= 33 e p= 0.001). After the rains individuals also Travel less, both alone (t= 3.105; n= 31 e p= 0,004) or in groups (t= 5.789; n= 33 e p= 0.001). Rest alone and in groups, and Social behavior is less frequent in period after the rains, but these are not statistically significant differences (z = -0.357; n= 31 e p = 0.721; z= -0.093; n= 33 e p = 0.926; z= -0.545; n= 33 e p = 0.586, respectively).

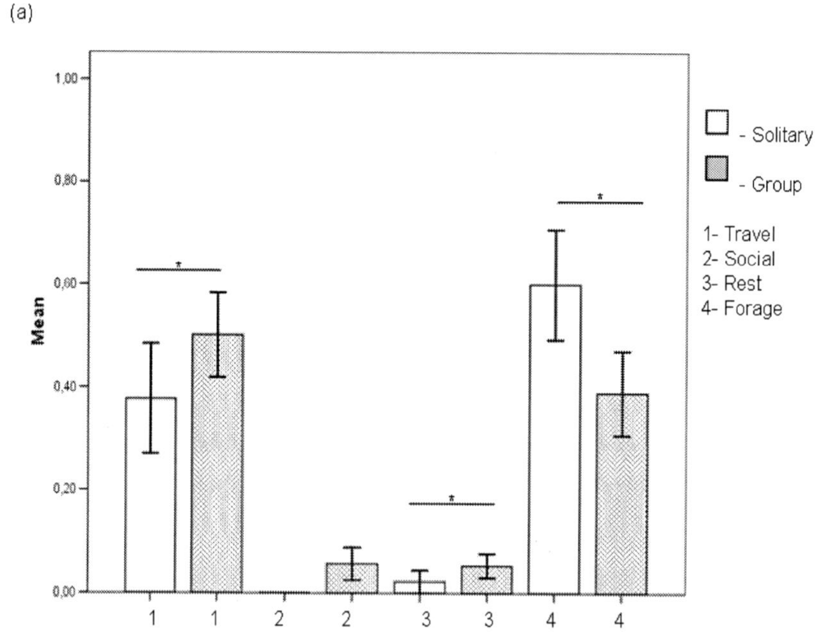

Figure Continued

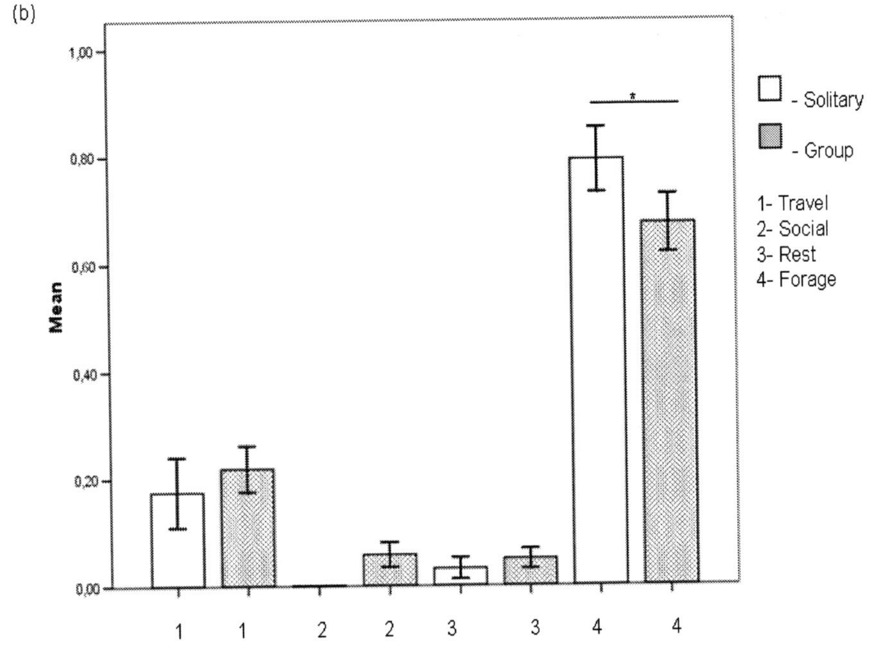

Figure 2. Time budget of adult *estuarine dolphins*, solitary and group situations. (a) period 1 - before the rains and (b) period 2 – after the rains (b). * = significative differences at p= 0.05.

However, when Forage behavior is decomposed in two components: search and persecution, it is possible to verify that the increase in forage behavior in period after the rains relates to an increase in search behavior but not an increase in actual persecution of preys (Persecution period 1: 11%; Persecution period 2: 13%; t= -1.669; n= 33 e p= 0.099; Search period 1: 30%; Search period 2: 54%; t= -5.501; n= 33 e p= 0.001).

CONCLUSION

For social species, studies of periods of solitary activity (e.g., during migration, during temporary detachment from groups) offer opportunities to observe individuals' activity budget without the constraint of group life, and offer glimpses on costs and benefits for group living. In this work we compared days of solitary activity and of group activity by adult estuarine dolphins. Although these animals tended to form small groups in the study area (less than five individuals- Tosi and Ferreira, 2008), the increase in Rest and Travel indicate that group life offers benefits in terms of protection from predators, similar to the main beneficial factor described in the specialized literature (Gnone et al., 2001).

Time employed in Forage decreases when individuals are in a group relative to when individuals are alone within the bay. There are two possible explanations for this result. Firstly, it is possible to argue for an increased foraging efficiency, probably related to a collective (cooperative?) hunting, as suggested by Hoelzel, 1991; Neumann and Orams, 2005 for other odontocetes. In fact, it is common to observe dyads and even groups of animals following fish schools towards shallow waters, and there are descriptions of Sotalia encircling fish schools (Silva and Best, 1996; Lodi and Hetzel, 1998; Geise et al., 1999). Conversely, it is possible to argue that there is an increase feeding competition in groups, with individuals having fewer opportunities for successful feeding when in groups, therefore, the decreased time in foraging observed.

The lack of differences between Rest and Travel solitary and group activities during the period after the rains may be related to an overall increase in Forage activity observed in this period. A change in behavior patterns during periods of less food abundance was also described in other areas occupied by estuarine dolphins (Daura-Jorge et al., 2004).

Despite this advantage of life in groups, estuarine dolphins tend to form small groups, with a modal size of less than 6 individuals in diverse populations (Geise et al. 1999; Lodi, 2003; Daura-Jorge et al., 2005; Azevedo et al., 2005; Queiroz, 2006). Estuarine dolphins present site fidelity, and the factors that hinder these animals from forming larger groups may be related to ecological constraints such as the availability of prey in each area (as suggested by Queiroz et al., sub.).

Based on Social Intelligence hypothesis, Connor (2007) has recently argued that large brains in odontocetes, or more specifically in bottlenose dolphins (Tursiops aduncus), derives from the need to keep stable alliances in fission-fusion societies, due to a risky environment. The type of social organisation of estuarine dolphins is still unknown. Monteiro-Filho (2000) suggested that this species forms monogamic pairs due to the high incidence of dyads and calves in a population inhabiting a protected bay (Cananéia). However, other studies reveal that groups of up to 40 individuals may split into dyads within foraging areas (Lodi, 2003; Daura-Jorge et al., 2005), although no study has directly related large group sizes to efficiency in foraging. In addition Tosi and Ferreira (sub.) discuss that estuarine dolphins do not present social displays (synchrony and petting) typical of the bottlenose population described by Connor (2007).

Protection from predators and foraging efficiency are two main ecological pressures suggested for group living, although species differ in which is strongest, and on the degree of dependency on group living. Although our analyses did not differentiate among various group sizes (which is indicative of social complexity), according to the data presented in this work it is possible to say that the advantages for groups relate to protection from predators (increases in Rest and Travel). However, data are more obscure regarding the advantages for collective hunt, an ecological pressure commonly suggested for group living in odontocetes considering their large size and their mobile prey (Mann et al., 2000). Considering that the study area is a bay with direct contact with the ocean, and about 7.5 km apart from the nearest river mouth, where large fish schools are occasionally but not often observed, it is possible to infer that living in a group for these dolphins is more related to protection from predators than to foraging efficiency. More detailed studies are needed to properly confirm this suggestion, but if this is the case, the sociality of these dolphins may not be as complex as that of Tursiops, and it is in accordance with the absence of elaborated social displays presented by the dolphins in this area.

ACKNOWLEDGMENTS

We would like to thank CAPES for financial support; Projeto Pequenos Cetáceos for logistical support in filed work and Lídio França do Nascimento for drawing figure 1. RGF was supported by Capes Grant PRODOC/ PRODEMA UFRN.

REFERENCES

Azevedo, A.F.; Viana, S.C.; Oliveira, A.M. and Sluys, M.V. (2005). Group characteristics of marine tucuxis (*Sotalia fluviatilis*) (Cetacean: Delphinidae) in Guanabara Bay, south eastern Brazil. *Journal Marine Biological, 85,* 209-212.

Balance, T. (1992). Habitat use patterns and ranges of the bottlenose dolphin in the Gulf of California, Mexico. *Marine Mammal Science, 8* (3): 262-274.

Byrne, R. and Whiten, A. (1988). *Machiavellian Intelligence: Social expertise and the evolution of intellect in monkeys, apes, and humans.* Oxford Science Publications.

Connor, R.C. (2007). Dolphin social intelligence: complex alliance relationships in bottlenose dolphins and a consideration of selective environments for extreme brain size evolution in mammals. *Philosophical Transactions of the Royal Society, B,*1-16.

Constantine R. ; Brunton D.H. and Dennis T. (2004). Dolphin-watching tour boats change bottlenose dolphin (*Tursiops truncatus*) behaviour. *Biological Conservation, 117,* 299-307.

Daura-Jorge, F.G.; Wedekin, L.L. and Simões-Lopes, P.C. (2004). Variação sazonal na intensidade dos deslocamentos do boto-cinza, *Sotalia guianensis* (Cetácea: Delphinidae), na Baía Norte da Ilha de Santa Catarina. *Biotemas, 17* (1): 203-216.

Daura-Jorge, F.G.; Wedekin, L.L.; Piacentini, V.Q. and Simões-Lopes, P.C. (2005). Seasonal and daily patterns of group size, cohesion and activity of the estuarine dolphin, *Sotalia guianensis* (P.J. Van Bénéden) (Cetecea, Delphinidae), in southern Brazil. *Revista Brasileira de Zoologia, 22* (4): 1014-1021.

Dunbar, R.I.M. (1988). *Primate Social Systems.* Croom Helm, London.

Dunbar, R.I.M. (1992). Neocortex size as a constrains on group size in primates. *Journal of Human Evolution, 20,* 469-493.

Geise, L.; Gomes, N. and Cerqueira, R. (1999). Behaviour, habitat use and population size of *Sotalia fluviatilis* (Gervais, 1853) (Cetacea: Delphinidae) in the Cananéia estuary region, São Paulo, Brazil. *Revista Brasileira de Biologia, 59,* 183-194.

Gnone, G.; Benoldi, C.; Bonsignore, B. and Fognani, P. (2001). Observations of rest behaviours in captive bottlenose dolphins (*Tursiops truncatus*). *Aquatic Mammals, 27* (1): 29-33.

Guilherme-Silveira, F.R. and Silva, F.J.L. (2007). Behavioural seasonality of the estuarine dolphin, *Sotalia guianensis*, on the north-eastern Brazilian coast. *JMBA2- Biodiversity Records,* Published on-line, 1-5.

Hoelzel, A.R. (1991). Killer whale predation on marine mammals at Punta Norte, Argentina; food sharing, provisioning and foraging strategy. *Behavioral Ecology and Sociobiology*, 29. 197-204.

Krebs, J.R. and Davies, N.B. (1996). *Introdução a ecologia comportamental.* 420 p.

Lee, P.C. (1994). *Social structure and evolution. In:* Slater, P.S.B. and Halliday, T.R. (eds.) *Behaviour and Evolution,* Cambridge University Press, Cambridge. p. 266-303.

Lodi, L. (2003). Tamanho e composição de grupo dos botos-cinza, *Sotalia guianensis* (van Benédén, 1864) (Cetácea, Delphinidae), na Baía de Paraty, Rio de Janeiro, Brasil. *Atlântica, 25,* 135-146.

Lodi, L. and Hetzel, B. (1998). Grandes agregações do boto-cinza (*Sotalia fluviatilis*) na Baía da Ilha Grande, Rio de Janeiro, *Revista Bioikos, 12* (2): 26-30.

Lusseau, D. (2003). Effects of Tour Boats on the Behavior of Bottlenose Dolphins: Using Markov Chains to Model Anthropogenic Impacts. *Conservation Biology, 17* (6): 1785-1793.

Mann, J. and Smuts, B. (1999). Behavioral development in wild bottlenose dolphin newborns (Tursiops sp.). *Behaviour, 136,* 529-566.

Mann, J.; Connor, R.; Tyack, P. and Whitehead, H. (2000). *Cetacean societies.* University of Chicago Press. 433p.

Monteiro-Filho, E.L.A. (2000). Group organization of the dolphin *Sotalia fluviatilis guianensis* in a estuary of seutheastern Brazil. *Ciencia e Cultura, 52* (2): 97-101.

Neumann, D.R. and Orams, M.B. (2005). *Behaviour and ecology of common dolphins (Delphinus delphis) and the impact of tourism in Mercury Bay, North Island, New Zealand. Science for conservation.* 254. Science and Technical Publishing Department of Conservation. Website: http://www.doc.govt.nz.

Norris, K.S. and Dohl, T.O. (1980). *The structure and functions of cetacean schools. In*: Herman, L M (ed) *cetacean behavior: Mechanisms and functions,* New york: Jolm Wiley and Sons.

Queiroz, R.E.M. (2006). *Estudos sobre orçamento de atividade do boto cinza (Sotalia guianensis) no litoral sul do Rio Grande do Norte.* Dissertação de Mestrado. Universidade Federal do Rio Grande do Norte, Natal, Brasil.

Seyfarth, R.M. (1977). A model of social grooming among adult female monkeys. *Journal Theoretical Biology, 65,* 671-698.

Silva, V.M.F. and Best, R.C. (1996). *Sotalia fluviatilis. Mammalian Species, 527,* 1-7.

Tosi, C.H. (2007). *Sincronismo comportamental em boto cinza (Sotalia guianensis).* Dissertação, Universidade Federal do Rio Grande do Norte, Natal, Brasil.

Van schaick , C. (1983). Why are diurnal primates living in groups? *Behaviour, 87,* 120-144.

Whiten, A and Byrne, R. (1997). *Machiavellian intelligence II – extensions and evaluations.* Cambridge University Pres, Cambridge.

Würsig, B. and Würsig, M. (1980). Behavior and ecology of the dusky dolphin, *Lagenorhynchus obscurus*, in the South Atlantic. *Fishery Bulletin, 77* (4): 871-890.

Zar, J. (1996). *Bioestatistical analyses* (3rd ed). Prentice Hall, New Jersey.

INDEX

J

K

L

M

N

O

P

R

S

T

U

V

W

Z